T0329927

Active Seismic Tomography

Active Seismic Tomography

Theory and Applications

Kalachand Sain
Wadia Institute of Himalayan Geology
Dehradun
Uttarakhand
India

Damodara Nara
CSIR-National Geophysical Research Institute
Hyderabad
India

For general information on our other products and services or for technical support, please contact our Customer Care Department within the United States at (800) 762-2974, outside the United States at (317) 572-3993 or fax (317) 572-4002.

Wiley also publishes its books in a variety of electronic formats. Some content that appears in print may not be available in electronic formats. For more information about Wiley products, visit our web site at www.wiley.com.

A catalogue record for this book is available from the Library of Congress

Hardback ISBN:9781119594864; epub ISBN:9781119594895; ePDF ISBN:9781119594833; oBook ISBN:9781119594925

Cover Image: © Raja Shoaib Turk/Shutterstock
Cover design by Wiley

Set in 9.5/12.5pt STIXTwoText by Integra Software Services Pvt. Ltd., Pondicherry, India

Contents

Preface

The seismic method is one of the most promising geophysical methods that can help to explore the Earth's interior. Numerous technology developments have been introduced in seismic methods over the decades to either understand or to use the full recorded seismic data from the field. Seismic tomography is a seismic imaging technique used to delineate the subsurface linked with structures and numerical values of different physical parameters such as the seismic velocity, density, etc.

Seismic tomography has grown from a very simple to a complex scenario, i.e., at first, we started to use traveltime data alone for seismic imaging in seismic tomography. Over the decades, seismic tomography has evolved to exploit comprehensive components such as traveltime, amplitudes, frequencies, etc. of the seismogram. This all happened because of tremendous advancements in computer technologies; however, successful applications of full waveform tomography are still very limited, and more understanding of challenges like cycle-skipping, selection of suitable norms, frequency bands, and implementation to 3D domains is needed.

This book aims to introduce active seismic tomography starting from traveltime tomography to full waveform tomography. It also covers both synthetic and real field data applications. It is divided into two sections. Section I is dedicated to theory. Chapter 1 introduces the pioneering history of seismic tomography along with developments occuring in the last two decades. Chapter 2 describes the derivation of wave equations and their different forms. Chapter 3 covers the forward problem of tomography and describes the methods for solving the seismic wave equation. Chapter 4 presents the inverse problem of tomography and describes its methods. Chapter 5 is dedicated to the subsurface parameterization to cover the variety of subsurface parameterizations in obtaining the best possible responses. This will include different types of parameterization, such as grids, cells, blocks, etc., and will discuss the pros and cons of each category for the proper selection that will lead to expected geological outcomes from the data without any bias.

Chapter 6 presents seismic traveltime tomography to discuss the theory of travel time tomography developed over the last two decades. This will summarize the advanced traveltime tomographic concepts scattered in research publications. The last chapter of section I, chapter 7, covers seismic full waveform tomography to summarize advanced concepts scattered in research articles.

Section II focuses on applying seismic tomography techniques for both traveltime and full waveform tomography. Chapter 8 is dedicated to applying seismic tomography to synthetic seismic data in understanding the role of several parameters involved in tomography for the best possible results. Chapter 9 covers the application of traveltime tomography to seismic data from the Kerala-Konkan offshore basin, western Indian margin. Chapter 10 is dedicated to the sophisticated full waveform tomography of seismic data to emphasize the importance of the technique over the traveltime tomographic approximation in obtaining high-resolution velocity models. Chapter 11 covers advanced seismic processing using tomographic results that showcase enhanced subsurface imaging using a tomographic velocity model. We have discussed the advanced seismic processing techniques like migration that needs proper velocity models to improve the image. We also summarize the future scope of the technique in chapter 12. This chapter covers the pros and cons of seismic tomography, hurdles, and a probable way to tackle the field data, future areas of research, etc. This chapter covers directions for future research in the emerging field of seismic tomography that may provide guidance to academicians and professionals for advancing research.

Since seismic tomography is one of the most prominent velocity-building techniques, we hope that this book will fill the gap among researchers, academicians, and explorationists in understanding the intricacies involved in seismic tomography for its successful applications to field data. We also believe that it will guide to young scientists in pursuing their careers in the frontier area of research.

We take this opportunity to convey our gratitude to our teachers and professors who nourished us from the beginning. The Director, CSIR-NGRI is acknowledged for according permission to publish this book. Mr. R.K. Srivastava, Director (Exploration), Oil & Natural Gas Corporation (ONGC) Ltd. is thanked profusely for his constant support of pursuing advanced research. We thank ONGC personnel Sri G.C. Katiyar, Mrs. Lata S. Pandurangi, Sri P.H. Mane and Mr. N. Chandrasekhar for their fruitful discussions from time to time and for providing seismic data. Generic Mapping Tools (Wessel et al. 2013) software was used for plotting some of the figures. Damodara Nara thanks to DST for providing the INSPIRE Faculty Fellowship.

Last but not the least, we gratefully acknowledge our families for their support, patience, and care, which enabled us to write this book. K.S. affectionately thanks his wife Tumpa and son Ritwik for their inspiration and love in this journey.

D.N. expresses sincere thanks to Dr. Kalachand Sain for his endless support and timely encouragement, which allowed for his participation in many national/ international conferences, course works, and R & D translational research works, which have enhanced his scientific temperament. This is a contribution to FTT research of CSIR-NGRI and GAP-822-28(ND).

Kalachand Sain
Wadia Institute of Himalayan Geology, India
Damodara Nara
CSIR-National Geophysical Research Institute, India

About the Authors

Kalachand Sain is the Director of the Wadia Institute of Himalayan Geology in Dehradun, India. Previously, he was the Chief Scientist at the CSIR-National Geophysical Research Institute (NGRI) in Hyderabad, India. He has an MSc (Tech) in Applied Geophysics from the IIT-Indian School of Mines, Dhanbad and a PhD in Active Source Seismology from CSIR-NGRI, Hyderabad, India. He spent time as a post-doctoral fellow at Cambridge University (UK) and Rice University (USA), and was a visiting scientist at the United States Geological Survey. His research interests include exploration of gas hydrates, imaging sub-volcanic sediments, understanding the evolution of sedimentary basins and earthquake processes, and providing geotectonic implications, including the Himalayan orogeny, and glaciological and landslides hazards. He has also built expertise in traveltime tomography, AVO modelling, full-waveform tomography, advanced processing, seismic attenuation and meta attributes, artificial intelligence, rock physics modelling, and interpretation of 2-D/3-D seismic data. He is a Fellow of all three Indian science academies and is the recipient of numerous medals and awards including the National Mineral Award, National Award of Excellence in Geosciences, J.C. Bose National Fellowship, Decennial Award & Anni Talwani Memorial Prize of Indian Geophysical Union, and Distinguished IIT-ISM Alumnus Award.

Dr. Damodara Nara is a scientist at CSIR-NGRI in Hyderabad, India. He received a MSc in Mathematics with Outstanding grade from Sri Venkatewara University, Tirupati and PhD in Science (Geophysics) from the Academy of Scientific and Innovative Research, Hyderabad. He has published several publications in SCI journals and in many national/international conferences. His research interests include seismic data processing, and seismic traveltime and full waveform tomography, and development of numerical algorithms for seismic data modelling and inversion. He earned the best poster award and the best young researcher award during the annual conventions of the Indian Geophysical Union and was awarded one of the prestigious Inspire faculty fellowships from the Department of Science & Technology, Govt. of India.

Glossary

Title	Definition
Acceleration	The rate of change of the velocity of an object defined as acceleration.
Acoustic	The branch of physics that deals with the study of mechanical waves in solids, gases, and liquids including topics such as vibration, sound, ultrasound, and infrasound.
Amplitude	The maximum distance/displacement moved by a particle/point from its equilibrium position on a vibrating body or wave measured.

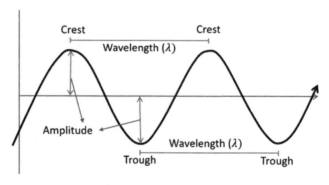

Title	Definition
Anisotropic	The mechanical properties of the material are not the same in different directions at an arbitrary point during its rotation of axes at a point.

Attenuation	The loss/reduction of energy of something over the propagation of it in the medium.

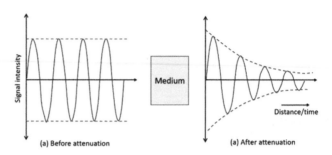

(a) Before attenuation   (a) After attenuation

Convolution	Convolution is a mathematical integral operation on two functions (f and g) that produces a third function (h) to expresses how the shape of one is modified or overlapped or shifted by the other.
Critical angle	The critical angle is the angle of incidence, for which the angle of refraction is 90°; i.e., It is an incident angle above which the total internal reflection occurs when light enters a denser medium from a comparatively rarer medium.
Data	The known direct measurements made by the observer from the real world with a specific objective.
Discretization	By which the computational domain is divided into so-called sub-domains or elements and their intersection points are defined as nodes.
Displacement	Displacement is the measurement of a straight path between two positions/objects, irrespective of the path travelled from one position/object to another. It is always a vector quantity.

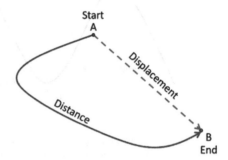

Distance	Distance is the measurement of a path travelled between two positions/objects. It is always scalar quantity. See figure in Displacement definition.

Domain	In general, the domain refers to the set of all inputs given to a function. In seismics, we parameterize the subsurface to process the inputs signal in either time/frequency forms, and these are characterized as domains in which we do further processes based on mathematical relations.
Elasticity	The deformations of the body are completely reversible, once the external forces or load is removed from the body.
Error	Error is defined as the measure of the estimated difference between the calculated value of a function and its true value.
Force	Force is an external influence that can change the motion of the body or the state of the rest position. It is a vector quantity.
Forward problem	The problem by which we can calculate the theoretical response for a particular model from a numerical representation.

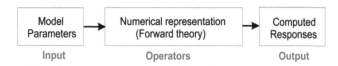

Global maxima	In Optimization problems, the function exhibits many extrema (minima/maxima) for the range of input, among those, the point at which the function having the largest value is termed as global maxima. See figure in *global minima*.
Global minima	In Optimization problems, the function exhibits many extrema (minima/maxima) for the range of input, among those, the point at which the function having the smallest value is termed as global minima.

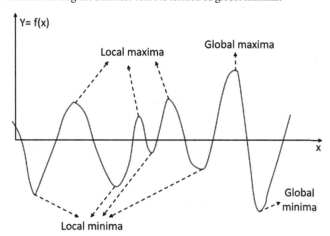

Ground roll	Ground roll is a type of seismic coherent noise caused by surface waves. It is characterized with high amplitudes and low frequency and low velocity in seismic records.

Homogeneous	Material property doesn't change during its translation.; i.e., at every point the material properties are invariant.
Heterogeneous	Material properties will change during their translation; i.e., the material properties vary from point to point.
Interpolation	Estimation or prediction of new data points from the range of available known data points is referred to as interpolation.
Inverse problem	The problem by which we can calculate the model parameter estimates of a particular model, which caused for observations from inverse theory. Note that the role of inverse theory is to provide information about unknown numerical parameters that go into the model, not to provide the model itself.

```
┌──────────────┐      ┌──────────────────────┐      ┌──────────────┐
│ Observational│─────▶│  Mathematical Tools  │─────▶│ Estimates of │
│    data      │      │   (Inverse Theory)   │      │    system    │
│              │      │                      │      │  parameters  │
└──────────────┘      └──────────────────────┘      └──────────────┘
     Input                   Operators                   Output
```

Isotropic	The mechanical properties of the material are the same in all directions at an arbitrary point during its axel rotation at a point.
First break/Direct	The earliest refracted arrival energy propagated from source to receiver associated with the weathering layer often constitutes a direct/first arrival on seismic shot gather. These are very useful in making statical corrections in the seismic data processing sequence.
Frequency	The number of repeated events occurring per unit of time is called the frequency.
Local Maxima	In Optimization problems, the function exhibits many extrema (minima/maxima) for the range of input, among those, the point at which the function having the maximum value is termed as maxima/local maxima. See figure in *global minima*.
Seismic Migration	The process to create an accurate image of the subsurface by geometrically re-locating the seismic events in either space or time to the actual location occurred in the subsurface rather than the location that it was recorded at the surface.
Local Minima	In Optimization problems, the function exhibits many extrema (minima/maxima) for the range of input, among those, the point at which the function having the minimum value is termed as minima/local minima. See figure in *global minima*.
Model	The defined relationship between the data and parameters is referred to as the model. It is used for calculating the unknown properties from the known measurements.
Multiples	The seismic events that undergo more than one reflection through the subsurface is characterized as multiples in seismic records.
Seismic Noise	The unwanted or un-interpretable components of seismic signal from recorded seismic data are characterized as seismic noise.

Numerical method	The method/tools used for solving the numerical problems. Some of the examples of popular numerical methods for geophysical applications are finite-difference and finite element methods.
Parameter	The unknowns that we want to answer in terms of numerical values of the specific properties of the world.
Time Period	The time taken for one cycle length is called the time period. The unit of time period is second.
Quality factor	The seismic energy dissipated over the time in various ways. The dissipated energy or attenuation of the energy measured in terms of dimensionless seismic quality factor (Q) stored in one cycle.
Ray	It is a perpendicular line drawn on a wavefront at any point in the direction of the wave propagation.

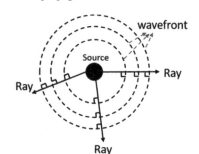

Receiver	The instruments used for recording the seismic signal is called the receiver. When used for acquisition in water, it is called the hydrophone, and when it is used for land seismic survey, it is called the geophone.
Resolution	Resolution measures of how large an object needs to be in order to be seen. It can be measured in two ways: in vertical and horizontal directions.
Scalar	The physical quantity used for measuring only the magnitude. Ex: Density, mass, time and volume.
Seismology	It is a branch of geophysics that is used to study the elastic/seismic waves propagated through the earth for imaging the subsurface.
Slowness	The inverse of the velocity (distance covered by an object in unit time) is defined as the slowness.
Source	The device that generates the propagated energy to perform the dynamic disturbance through the medium. In seismics, there are different kinds of sources available to generate energy: hammer, explosives, vibrator, Airgun etc. along with the natural source in the form of earthquake.
Stacking	In seismics, stacking refers to adding seismic traces into a single trace for improving the data quality and reduce the noise.

Strain	The strain is defined as the amount of deformation experienced by the body in the direction of the force applied.
Stress	The stress is defined as force per unit area.
Traveltime	The time to reach the energy or wave from the seismic source to the receiver is called traveltime.
Tomography	It is type of inverse problem by which we determine the model structure by back projecting the data along a path connecting a source and a receiver.
Vector	The physical quantity used for measuring both the magnitude and direction. Ex: Force, displacement, acceleration.
Velocity	The rate of change in object position as measured in a particular standard of time is called velocity.
Wave	The dynamic disturbance caused by an energy source and that is propagated through a medium is defined as waves. In seismics, we have two categories: body waves (P waves and S waves) and surface waves (Love waves and Rayleigh waves).
Wavefront	It is representation of all parts of the wave that are in same phase. The shape of the wavefront is either spherical or circular depending on the nature of the source that emit the energy. See the figure in *Ray*.
Wavelength	The distance between two successive crests or troughs in wave propagation is defined as the wavelength. See the figure in the *Amplitude* definition.
Wavenumber	The Spatial frequency of the wave is referred as the wavenumber.

Section I

Theory

1

Pioneering History

1.1 Introduction

Of all geophysical methods, the seismic method is the most promising one in terms of revealing the internal structure and composition of the Earth from a few shallow meters to a deeper few tens of kilometers. The delineation of the subsurface from the seismic method always depends on the type of seismic data acquisitions, which follows recording different seismic waves by a variety of geometries. The wide range of geometries includes seismic reflection, refraction, wide-angle seismic reflection, etc. as part of active-source seismic exploration and engineering applications and for local, global, and tele-seismic, etc. applications as a part of passive seismology. The low-angle seismic reflection data provides better structural information but lacks accurate velocity derivation, whereas the refraction/ wide-angle reflection data provides better velocity information but with poor structural resolution of the crust and uppermost mantle. Thus, co-incident seismic reflection and refraction/wide-angle reflection experiments provide an accurate velocity-structure of a region. If the region is complex, such as is the thrust fold belt area or basalt covered province, it is rather difficult to image the subsurface by conventionally processing multi-fold seismic reflection data. Even state-of-the-art, pre-stack depth migration (PSDM) cannot be applied to the near-vertical reflection data without proper velocity information. On the other hand, seismic refraction/ wide-angle reflection or ocean bottom seismic data can provide reliable velocity imaging of the subsurface even in a difficult terrain. The traveltimes of wide-angle seismic wave fields, recorded at the surface, are modeled or inverted through forward modeling or inversion theory to produce large-wavelength variations of seismic velocity structure, which can be broadly used for geo-tectonic implications or as an input for pre-stack depth or for reverse time migration for fine tuning subsurface images. Aki and Lee (1976) introduced the technique called seismic tomography, which is initially borrowed from medical imaging. It was used to reconstruct

Active Seismic Tomography: Theory and Applications, First Edition. Kalachand Sain and Damodara Nara.
© 2023 John Wiley & Sons, Inc. Published 2023 by John Wiley & Sons, Inc.

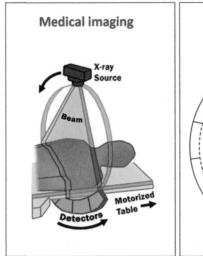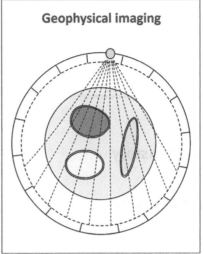

Figure 1.1 Illustration of medical and geophysical imaging techniques. Left: The emitted X-rays travel through the human body and are detected by detectors; this will be processed to reconstruct bone density using X-rays absorption in medical imaging. Right: The generated seismic waves travel through the Earth and are recorded at different receiver locations surrounded by the Earth. The recorded data will be processed to reconstruct the seismic velocity in geophysical imaging.

the human body's density information through X-ray emissions and density-related absorption (Figure 1.1). The same principles are utilized in geophysical imaging to delineate the velocity from the emission of seismic waves and their components (traveltimes, amplitudes, frequencies, phases, etc.) related to velocity by solving it as an inverse problem (Figure 1.1). The term tomography is derived from Greek i.e., Tomography = tomo + graphy = slice + picture/section.

1.2 Applications

Seismic data contains one of the most valuable pieces of information for investigating the Earth's internal structure and composition. A number of methods exists for extracting subsurface structure from seismic data with different objectives in a wide range of geophysical applications (Figure 1.2). Here, we focus on the applications of seismic tomography using different components (traveltimes, amplitudes, frequencies, phases, etc.) of seismic data records. First and foremost, commonly known researchers Aki and Lee (1976) used the seismic tomography technique for local earthquakes delineate the 3D velocity structure beneath California from local earthquakes. The traveltime data was collected by 60 stations from 32 local

Figure 1.2 The wide range of applications of seismic tomography includes cross-well studies for petroleum exploration, global studies for tectonic activities, earthquake risks studies for upper crustal faults and basin geometries in understanding geo-tectonics of the region and surface-surface studies for hydrocarbon exploration.

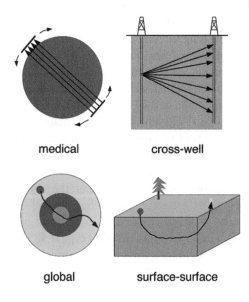

earthquakes. The data parameterized the subsurface into 264 constant slowness blocks and assumed that the inversion was linear because ray paths were assumed as a straight line through the homogeneous initial model. Later many researchers started to use the seismic data either in part or in full for the recorded seismograms for tomography; though it was named differently as traveltime, diffraction, full waveform tomography, etc. However, as traveltime tomography becomes conventional in delineating long-wavelength velocity structures, the other one, full waveform tomography, still challenges the community.

Bois et al. (1972) also implemented the technique for well-to-well recorded data, and their approach used a regular grid of nodes for the parameterization of the subsurface between two wells. In the late '80s and during the '90s, there were numerous studies developed in both raytracing and in wave equation modeling methods. Among those, the most popular studies were by Zelt and Smith (1992) and Zelt and Barton (1998), which used traveltimes by solving the eikonal equation in forward problems and in using the least square inversion methods to minimize the objective function. Zelt and Smith (1992) dealt with both reflection and refractions whereas Zelt and Barton (1998) dealt with refracted or diving, direct and diffracted phases alone except reflection traveltime data to develop the subsurface velocity structure. Taillandier et al. (2009) introduced the adjoint state method for using gradient calculation on traveltime tomography to avoid the ray tracing and estimation characteristic of the Fréchet derivative matrix. Lelièvre et al. (2011) introduced the fast marching method to avoid ray tracing the first arrival traveltime inversion by calculating the sensitivity of information through an explicit symbolic differentiation of the forward modeling equations and by calculating directly during the forward solution. Huang and Bellefleur (2012) implemented the fast-sweeping method and adjoint-state method for joint inversion of both transmission and reflection traveltimes. The adjoint-state method required the two forward modeling solutions and thus circumvented the computation of the Fréchet derivative matrix estimation. This method also facilitates that the gradient can be calculated independently for each shot. Waheed and Alkhalifah (2015) extended the adjoint-state method to anisotropic media by taking advantage of the anisotropic eikonal solvers developed by Waheed et al. (2014, 2015). Afterward, Waheed et al. (2016) then introduced anisotropic first arrival traveltime tomography based on the adjoint-state method. On the other hand, in the early 1980s, the researchers started to develop numerical solutions for the wave equation for seismic exploration problems (Bamberger et al. (1982) and Tarantola (1984)). This helps exploit all types of seismic waves in synthetic seismograms to image the Earth's subsurface and is coined as full waveform tomography. Pratt and Worthington (1988), Pratt and Goulty (1991), Geller and Hara (1993), Song et al. (1995), Pratt and Shipp (1999), Pratt et al. (1998), and Pratt (1999) applied the full waveform tomography in frequency-domain, and Kolb et al. (1986) and Gauthier et al. (1986) applied in time-domain full waveform tomography. Further, Shin and Cha (2008) suggested an alternative Laplace-domain waveform inversion to sensitivity toward the initial model, which lacks in low frequency components. The wide-angle full waveform tomography is the main topic in advanced seismic imaging techniques, and the studies of Brenders and Pratt (2007a, 2007b) and Operto et al. (2006) from can be treated as benchmark.

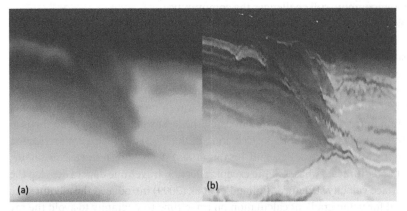

(a)

(b)

Figure 1.3 The sample derived velocity structure from (a) Traveltime tomography and (b) Full waveform tomography. (Kapoor et al., 2013).

However, we have very limited successful full waveform tomography applications within the field of seismic data due to various aspects like initial model, non-linearity of the problem, lack of low-frequencies, computational costs, etc. Alkhalifah (2015a, 2015b) analyzed the influence of scattering angles to the gradient model that propagates to update the model; we need to be cautious in the selection of scattering angles. Over the decades, tomography is a conventional method either by traveltimes alone or by full waveform for purposes of exploratively imaging the velocity model. Ray-based traveltime tomography gives the long wavelength structure, and this can be used as an initial model for full waveform-based techniques to delineate the sub wavelength structure of the subsurface (Figure 1.3).

1.3 Terminology

Data: The known direct measurements made by the observer from the real world with a specific objective.

Parameter: The unknowns that we want to answer in terms of numerical values of the specific properties of the world.

Model: The defined relationship between the data and parameters is referred to as the model. It is used for calculating the unknown properties from the known measurements of the world.

Tomography: It is a type inverse problem by which we determine the model structure by means of back projecting data along a path that connects a source and a receiver.

2

Seismic Wave Equation

2.1   Elastodynamic Wave Equations

The theory of elasticity relates the applied forces to the resulting changes in body size and shape. To understand the different types of seismic waves that propagate through the Earth away from a seismic source, some elementary concepts of stress and strain need to be considered (Kearey et al. 2002). "Stress" is defined as force per unit area. There are two types of forces that can act on an object: body force and surface force. Body force acts everywhere within an object and the resulting net body force is proportional to the volume of the object. A surface force acts on the surface of an object and the net surface force is proportional to the surface area of the object. "Strain" is deformation measured as the fractional change in dimension or as volume induced by stress. Strain is a dimensionless quantity.

Generally, we formulate wave equations based on some assumptions to either simplify the problem or to understand different material properties for which the analysis is valid, and some these assumptions are listed. Thus, it suggested the reader know these assumptions before analyzing any material.

Linearity: We normally assume two types of linearities: material linearity and Geometric linearity. Material linearity tells the relationship between stress-strain as Hooke's law and the Geometric linearity, which gives the deformations or small strains among the materials.

Elastic: The deformations of the body are completely reversible once the external forces or load is removed from body.

Continuum: The Matter of the any material is continuously distributed for all size scales without any voids.

Homogeneous: Material properties don't change during their translation; i.e., at every point material properties are invariant.

Active Seismic Tomography: Theory and Applications, First Edition. Kalachand Sain and Damodara Nara.

Inhomogeneous or Heterogeneous: Material properties will change during their translation; i.e., material properties vary from point to point.

Isotropic: The mechanical properties of the materials are the same in all directions at an arbitrary point during their rotation of axes at a point.

Anisotropic: The mechanical properties of the materials that are not same in different directions at an arbitrary point during their rotation of axes at a point.

Consider that in a continuous medium, a region of matter volume V surrounded *b* is a closed surface S (Figure 2.1). We can define the stress tensor for each point in terms of volume with nine stress components to keep the body in equilibrium as follows:

$$\left\{ \sigma_{ij} \right\} = \begin{pmatrix} \sigma_{xx} & \sigma_{xy} & \sigma_{xz} \\ \sigma_{yx} & \sigma_{yy} & \sigma_{yz} \\ \sigma_{zx} & \sigma_{zy} & \sigma_{zz} \end{pmatrix} \tag{2.1}$$

The stress acting upon one surface composed into three components as one (σ_{xx}) is normal for the surface on which it acts and on which other two (σ_{xy} and σ_{xz}) are tangential to the surface. In the subscript, first and second refer to being normal and tangential to the surface, respectively.

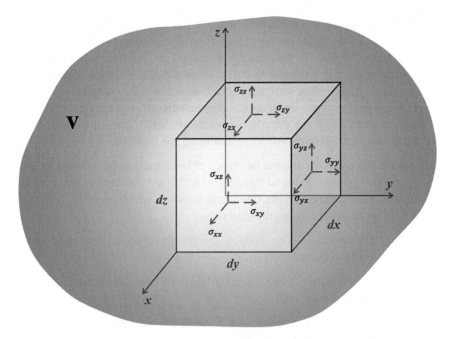

Figure 2.1 Stress components acting on a small volume surrounding a point within an elastic solid continuous medium of volume V.

Once the stress is applied to the volume body, it induces deformation, which vanishes once it relaxes. Thus, we can define the associated strain tensor for any two points that are displaced to new positions in a body as

$$\left\{e_{ij}\right\} = \begin{pmatrix} e_{xx} & e_{xy} & e_{xz} \\ e_{yx} & e_{yy} & e_{yz} \\ e_{zx} & e_{zy} & e_{zz} \end{pmatrix} \tag{2.2}$$

We now define the general expression for the net force acting on the object of volume V in all three directions using the stress tensor as

$$F_i = \sum_{j=1}^{3} \frac{\partial \sigma_{ji}}{\partial x_j} dVi, \, j = 1,2,3 \tag{2.3}$$

As we know, Newton's second law states the force that acts on an object is equal to the product of mass (m) and its acceleration. Let's consider u to be a small displacement of the small volume element dv due to the waves passing through an object of volume V.

$$Force(F) = mass(m) X \; acceleration(a)$$

$$F = (\rho dV) X \left(\frac{\partial^2 u}{\partial t^2} \right) = \rho \frac{\partial^2 u}{\partial t^2} dV$$

This can be written in a generalized form for all three directions as

$$F_i = \rho \frac{\partial^2 u_i}{\partial t^2} dVi = 1,2,3 \tag{2.4}$$

Considering the volume force fdV and from Equations (2.3) and (2.4), we have

$$\rho \frac{\partial^2 u_i}{\partial t^2} dV = \sum_{j=1}^{3} \frac{\partial \sigma_{ji}}{\partial x_j} dV + f_i dV$$

Divided with dV, the above equation reduces the elastodynamic equation form to

$$\rho \frac{\partial^2 u_i}{\partial t^2} = \sum_{j=1}^{3} \frac{\partial \sigma_{ji}}{\partial x_j} + f_i \tag{2.5}$$

Depending on the properties of the materials, we can formulate a different set of equations. From Hooke's law, we can establish a stress-strain relationship for elastic solids, and we can state that the strain is directly proportional to the applied stress. This can be written as

$$\sigma_{ji} = \lambda\Delta\delta_{ij} + 2\mu\epsilon_{ji} \tag{2.6}$$

$$\epsilon_{ij} = \frac{1}{2}\left(\frac{\partial u_i}{\partial x_j} + \frac{\partial u_j}{\partial x_i}\right)$$

Where δ_{ij} is the Kronecker delta, λ *and* μ are the Lame parameters and independent of time, ϵ_{ij} is the strain tensor. The set of Equations (2.5) and (2.6) represents the Displacement-Stress form of elastodynamic equations. Using $v_i = \dfrac{\partial u_i}{\partial t}$ and taking the time derivative for Equation (2.6) as

$$\frac{\partial \sigma j_i}{\partial t} = \lambda\frac{\partial\Delta}{\partial t}\delta_{ij} + 2\mu\frac{\partial\epsilon_{ji}}{\partial t} \tag{2.7}$$

$$\frac{\partial\epsilon_{ji}}{\partial t} = \frac{1}{2}\left(\frac{\partial v_i}{\partial x_j} + \frac{\partial v_j}{\partial x_i}\right)$$

And with Equation (2.3) re-written as

$$\rho\frac{\partial v_i}{\partial t} = \frac{\partial\sigma_{ji}}{\partial x_j} + f_i \tag{2.8}$$

Now these set of Equations (2.7) and (2.8) combinedly form a Stress-Velocity for motion.

2.2 Acoustic Wave Equation

The set of elastodynamic equations derived in section 2.1 is only considered when the medium is solid and when it describes both body and surface waves. If we consider the medium is a fluid/gas, then we have only compressional waves and no shear waves. So, we can derive another form of the equation of motion called the acoustic/scalar wave equation by taking the constant density and zero value of shear stress with ($\sigma_{ji} = 0$) and what's called the pressure field. The stress tensor becomes $\sigma_{ji} = -P\delta_{ij}$, and the acoustic equation is defined as

$$\frac{\partial^2 P}{\partial t^2} = c^2\nabla^2 P - c^2\nabla.\mathbf{f} \tag{2.9}$$

Where c defines the acoustic velocity as $c = \sqrt{\frac{\kappa}{\rho}}$, κ, which is bulk modulus.

2.3   Boundary Conditions

The equations of motion are partial differential equations, and these are always confined by initial and boundary conditions. Initially, we will not observe any system changes in terms of either displacement or velocity before energy is applied. Energy is released only after the time t > 0. In most of the geophysical problems, the desired study area (generally the computational boundary in modeling) is bounded with the free surface at the top and with the remaining three boundaries at the left, right and bottom boundaries. These computational boundaries can generate the artificial reflections that backpropogate into the desired domain/interior region (Figure 2.2), and these contaminate the data. So, to mitigate the boundary reflections, we can implement the conditions, the so-called boundary conditions. There are different kinds of conditions, which includes the free surface boundary condition, which is always put on the top of model, absorbing the boundary condition (Clayton and Enquist 1977), and the Dirichlet and the perfectly matched layer (Berenger 1994 and Collino and Tsogka 2001), at the other sides. These are categorized into absorbing boundary conditions and boundary layers. The absorbing boundary conditions are mainly derived from paraxial approximations of the wave equation; thus, these always depend strongly on incident angles. On the other hand, the boundary layers introduce a thin layer along the artificial boundary by modifying the wave equation that follows the amplitude's rapid decay through a thin layer (Cerjan et al. 1985). This tapering has improved by another layer called the perfectly matched layer method, which considers the amplitude's exponential decay (Berenger 1994). The simple illustration of effects of different conditions are shown in Figure 2.3.

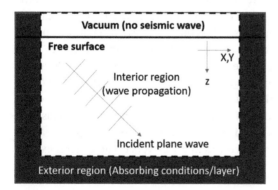

Figure 2.2   Illustration of the absorbing boundaries concept. The desired computational region is surrounded by the absorbing boundary.

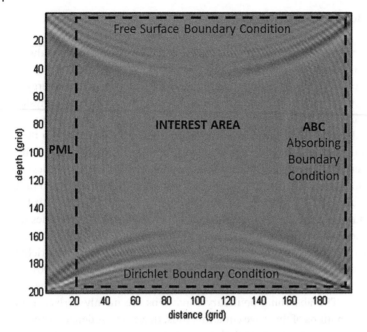

Figure 2.3 Sample illustration of four different boundary conditions Hardi and Sanny (2016).

2.4 Attenuation

In general, when the wave passes through the Earth from the source to receiver, the total energy applied at the source is different from the energy received at the receiver. This loss of energy happened due to geometrical or spherical spreading, to scattering and absorption properties. When the wave loses its energy by propagating radially from the source to receiver, the amplitude decreases with increasing distance; this is termed as geometrical spreading. The energy decay per unit wavefront is $1/r^2$ and the amplitude decay is $1/r$, where r is the radius of the wavefront. The attenuation is also evident in that the incident energy travels as reflected, refracted and diffracted at a reflector position. The energy loss is analyzed with the quality factor (Q). Q and attenuation are inversely proportional; for high Q values, the energy is attenuated less and vice-versa. The general expression is represented as

$$\frac{1}{Q} = -\frac{\Delta E}{2\pi E}$$

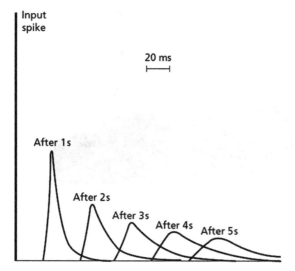

Figure 2.4 The progressive loss of energy for the input spike pulse signal travelled through the Earth over time (Anstey 1977).

Where E is total energy and ΔE is the loss of energy per propagated wave cycle.

For Homogeneous materials, the attenuation coefficients increase with increasing frequency. Thus, the higher frequencies are attenuated more rapidly than lower frequencies as a function of time or distance. We can observe these effects over the input spike signal case as shown in Figure 2.4.

2.5   Sources

The seismic experiment always depends on the energy generated from the source to ensure there is sufficient penetration depth, so the image of the subsurface has enough resolution. This source should have sufficient energy by which it can pro-duce a measurable signal with good signal-to-noise ratio. Different kinds of sources exist for generating seismic energy, and these are characterized into three types: impulsive, impact and vibrator. These includes hammer, Explosives, vibriosis, airgun, and earthquakes (for passive semiology), etc. Representative sample images of different sources are shown in figure 2.5. Based on the objective of the study and the medium, we can characterize the usage of different sources. In general, except for all airgun sources that are utilized for land acquisition

Figure 2.5 Sample images of the seismic sources used for exploration studies.

surveys, the airgun is used to generate pressure in offshore seismic data acquisition surveys. Once the energy is released from the source and is propagated through the Earth, we record the data with instruments at specific locations called receivers. These are generally, geophones, hydrophones, and seismometers.

3

Forward Problem of Tomography

3.1 Introduction

In this chapter, we describe the methods for modeling seismic wave propagation through the medium, i.e., Earth. In general, analytical methods don't afford solutions for realistic models of the Earth's subsurface through the equation of motion. On the other hand, numerical methods provide approximate solutions to the equation of motion for any complex simulated models of the Earth's interior. In numerical methods, the approximations used for spatial derivatives lead to a system of equations, which can be solved with linear algebra. These methods include numerous numerical methods depending on the discretization of space and time. Mainly the spatial discretization discriminates each method in the process of transforming the wave equation into the algebraic system. There are different classes of approaches including ray tracing methods, direct methods, and integral equation methods, which exist to solve the wave equation (Carcione et al. 2002). Each method has its own advantages and disadvantages depending on the objective of the problem in developing the algorithms for seismic wave propagation. Here we introduce the most widely used methods for most of the geophysical problems in a nutshell.

3.2 Finite-Difference Method

Finite-difference method (FDM) is one of the easiest methods available among the many methods that work to efficiently obtain wave equation solutions in both homogeneous and heterogeneous mediums. In FDM, the subsurface is parameterized with grids, and the parameter values are defined at grid positions. The FDM converts the partial differential equation (PDE) into matrix equations using approximations for the derivative at a grid position that is evaluated at the finite number of neighboring grids. The illustration representing the different approximations are

Active Seismic Tomography: Theory and Applications, First Edition. Kalachand Sain and Damodara Nara.
© 2023 John Wiley & Sons, Inc. Published 2023 by John Wiley & Sons, Inc.

shown in Figure 3.1. These equations form linear systems over discrete basis elements, and the result is usually a sparse matrix. This also allows us to investigate the order of accuracy of the approximations. Matrix-based algorithms represent matrices explicitly, and Matrix-free algorithms implicitly represent matrix values.

The difference approximations are constructed from the truncated Taylor expansions.

Taylor's expansion of the function $f(x)$ at $x + \Delta x$ and $x - \Delta x$ are as follows:

$$f(x+\Delta x)=f(x)+f'(x)(\Delta x)+\frac{f''(x)}{2!}(\Delta x)^2$$
$$+\frac{f'''(x)}{3!}(\Delta x)^3+\dots+\frac{f^n(x)}{n!}(\Delta x)^n+\dots \qquad (3.1)$$

$$f(x-\Delta x)=f(x)-f'(x)(\Delta x)+\frac{f''(x)}{2!}(\Delta x)^2$$
$$-\frac{f'''(x)}{3!}(\Delta x)^3+\dots+(-1)^n\frac{f^n(x)}{n!}(\Delta x)^n+\dots \qquad (3.2)$$

From Equation (3.1), we have

$$f'(x)=\frac{\left[f\left(x+\Delta x\right)-f\left(x\right)\right]}{\Delta x}-O(\Delta x)$$

$$f'(x)\approx\frac{\left[f\left(x+\Delta x\right)-f\left(x\right)\right]}{\Delta x}$$

Which defines the first-order approximation to a derivative, since the leading term of the approximation error is proportional to Δx and is simply called the

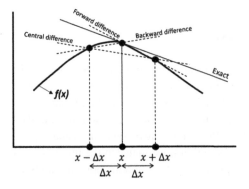

Figure 3.1 Illustrative example of different approximations to replace the partial derivatives of function.

forward-difference formula. The approximation error is called the *truncation error* and is, defined as the difference between the true derivative and the finite-difference approximation to it. This error is due to neglecting the higher-order terms of the Taylor series in finite-difference approximations.

Similarly, from Equation (3.2), the *first-order backward-difference formula* is written as follows.

$$f'(x) = \frac{\left[f(x) - f(x - \Delta x) \right]}{\Delta x} + O(\Delta x)$$

Again, subtracting Equation (3.2) from (3.1), we have

$$f'(x) = \frac{\left[f(x + \Delta x) - f(x - \Delta x) \right]}{2\Delta x} - O((\Delta x)^2)$$

Which defines the second-order accurate approximation of the first derivative and is called the *central-difference formula*.

And by summing Equations (3.1) and (3.2), we have

$$f''(x) = \frac{\left[f(x + \Delta x) - 2f(x) + f(x - \Delta x) \right]}{(\Delta x)^2} - O((\Delta x)^2)$$

Which is called the *accurate second-order approximation of the second derivative*. These difference equations are used to convert the differential equation into algebraic equations. Here we have given only a brief introduction to FDM. The application of FDM for seismic wave propagation modeling can be obtained from comprehensive documentation by Moczo et al. (2007, 2014). The finite-difference method is implemented in both uniform and non-uniform grids in the Cartesian system of coordinates. Further, the function values can approximate at the same grid positions in conventional grids and in different positions in either partly or fully staggered grids. This all aims to reduce memory requirements and to improve accuracy in the process of dealing with anisotropic media (Iturrarán-Viveros and Sánchez-Sesma 2011).

Initially, Alterman and Karal (1968) introduced the FDM to elastic wave propagation and Alford et al. (1974) to acoustic wave propagation along with the method's accuracy. Later Boore (1970) used and developed the snapshots of the seismic wave fields. Although, FDM has efficiently developed new algorithms, it has always parallelly linked with computational hardware development to introduce realistic applications in seismic imaging. The initial developments are only used to approximate all functions at the same grid positions. It was extended to define

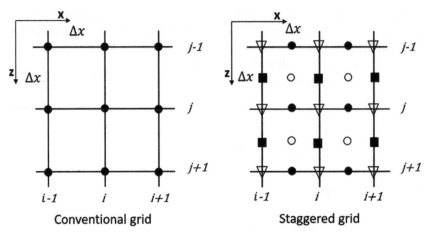

Figure 3.2 Simple illustration of different finite-difference grids.

the model parameters over the staggered grid like velocity/displacement at one grid whereas the stress tensor components are at different grid position (Madariaga 1976; Virieux and Madariaga 1982) as illustrated in Figure 3.2. This staggered grid parameterization received a wide range of positive applications to image the subsurface. Over the course of FDM development, different concepts like boundary conditions and simulation of non-uniform topographies were introduced for accurately simulating wave propagation even through heterogeneous media.

3.3 Finite-Element Method

At very first, the finite-difference methods are aimed to implement on regular grid (Figure 3.3) parameterizations for solving motion equations on simple geometries; however, it also extended to complex geological terrains, but it is very difficult to implement. To model complex geometries, another numerical method so-called the finite-element method (FEM) was introduced from structural engineering. In this, we parameterize the subsurface with numerous finite-elements (Figure 3.3), and each element is defined with single identical property. Finally, we assemble the parameterized finite number of elements into single units to make a complete system. This allows to modeler to define any complex geometries like realistic topography rather than one that only references a flat surface at the top; one can also feasibly obtain intermediate complex subsurface structures. The finite-element is generally defined with linear basis functions, and we solve the basis function to obtain the displacement/velocity field over the individual element; later we take the summation to obtain the total displacement.

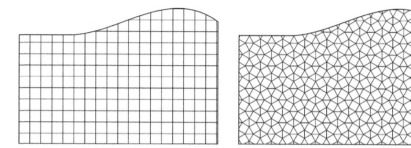

Figure 3.3 Representation of different parameterizations are used in numerical methods. Left: Regular grids and Right: Trinaular/irregular grid patterns.

$$u = \sum_{i=1}^{N} u_i(t)\varphi_i(x)$$

The displacement fields u_i are defined at the node points x_i and to the interim related to the coefficients of basis functions φ_i.

In general, one can define the basis functions for the different kinds of grids like triangles, squares, or curvilinear polygons. Mostly, FEM is applied for continuous functions and boundary value problems; i.e., we define the piecewise polynomials for a continuous function over grids to approximate the value and to overcome the difficulty in computing the exact functions. The following basic concepts need to be considered to apply the finite-element method.

Discretization: By which the computational domain divided into so-called subdomains or elements and their intersection points are defined as nodes.

Derivation of basis functions/ element equations: The variational formulation is necessary for an element. These are useful to define the element matrices.

Assembly: We achieve the full approximate solution by assembling the element equations that can be written in matrix form.

Boundary conditions: Like all other numerical methods, FEM also takes care to obtain the solutions without contaminating the solutions at the boundary region of the computing domain.

Solution of the equation: Once we implement the boundary conditions, we can solve the equation by any other standard technique, like LU decomposition.

By considering all of these, however, different varieties of FEMs exist, the Spectral element method and discontinuous Gelerkin method are most widely used for seismic modeling. The extensive details of the finite-element method can be found in popular classic books by Strang and Fix (1988), Zienkiewicz and Taylor (1989), and Zienkiewicz et al. (2013).

4

Inverse Problem

Tomography

4.1   Introduction

The wave equation solution from numerical methods produced the synthetic data in the forward problem. This solution is obtained by taking the initial assumption that the subsurface is beneficial and useful for solving the inverse problem in seismic tomography. This inverse problem will provide information for the initial model parameters that travel to obtain the next updated model till achieving the good match/minimum error between the synthetic and field seismic data; i.e., the role of inverse theory is to provide information about unknown numerical parameters that go into the initial/previous model, not to provide the model itself. One of the main difficulties from tomographic inversion is the non-uniqueness of the solution. The generalized tomographic inversion workflow is shown in Figure 4.1. Obtaining the differences between the synthetic data and the observed data is one of the first steps in the workflow, which can be measured with different norms; however, we mostly use the L_2 norm. The quantification of the differences is expressed in terms of the misfit functional represented with χ. During the tomographic inversion, we always try to find the optimum model that gives the global minimum of the misfit functional. This always a challenging task for to various reasons such as dealing with a large number of model parameters and the non-linearity of the model; thus, we adopt methods such as conjugate gradients or Newton to proceed the minimization iteratively based on gradient calculation. In general, the simplest linear inverse problem is represented with the matrix equation d = Gm. Here, the matrices d, m, and G are data, model, and data kernel respectively. This can be written in different forms based on the scope of discrete or continuous theory as follows:

$$\textit{Discrete inverse thoery}: d_i = \sum_{j=1}^{M} G_{ij} m_j$$

Active Seismic Tomography: Theory and Applications, First Edition. Kalachand Sain and Damodara Nara.

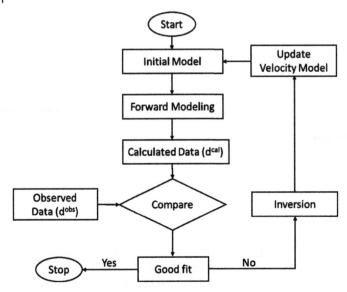

Figure 4.1 Generalized iterative workflow for seismic tomographic problem.

Continuous inverse theory : $d_i = \int G_i(x) m(x) dx$

Integral equation theory : $d(y) = \int G(y, x) m(x) dx$

Here x is an independent variable and subscript represents the number of data points and model parameters.

Since any geophysical data always has the data error in any form such as instrument, reading, computational round off error, noise, etc., we always approximate the solutions for the inverse problem in different forms. These can be viewed as estimates, bounding values, probability distributions, and weighted averages. The inverse problem was also characterized differently depending on the availability of the data points and of the required estimation of the number of model parameters. The inverse problem is said to be under-determined in the case of lacking data points rather than unknown model parameters. If we have more data points than unknown model parameters, then the problem is said to be an overdetermined case, i.e., the inverse problem does not provide the unique solution for the model parameters by having either less data or more data for under-determined and over determined problems, respectively. In overdetermined cases, we may overestimate the model parameters needed to be cautious. In addition to these two forms, we also have mixed and even determined cases for the inverse problem.

4.2   Inverse Problem

Consider we have an observed data d recorded at n receiver locations, and the suitable parameterized starting model m_o, which is close enough to the global optimum solution to allow successive linearizations without trapping in local minima. u is the predicted data using any forward solution of the wave equation at the same n receiver locations through the model m_o; therefore, we can define the data residual Δd as

$$\Delta d = \Delta d_i = d_i - u_i, i = 1, 2, 3, ..., n \text{ represents the receiver number} \quad (4.1)$$

From the weighted least-squares inversion, we have the objective function as

$$C(m) = \Delta d^t W_d \Delta d^* \quad (4.2)$$

Where Δd is the data residual matrix, the superscript t denotes the matrix transpose, the subscript * denotes the adjoint matrix, i.e., the transpose of the complex conjugate, and W_d denotes the data weighting operator that scales the relative contribution of each component of the matrix Δd during the inversion. To obtain the optimum approximate solution for the objective function, we use the iterative schemes in which the current model is updated with the preceding model so that the residual is the minimum for the updated model from the previous step. The generalized iterative equation is

$$m_i = m_{i-1} + \alpha_{i-1} h_{i-1} \text{ such that } C(m_i) < C(m_{i-1})$$

The model is updated in the descent direction (h_{i-1}) with quantity so-called step length $(\alpha_{i-1} > 0)$. This varies in each iterative scheme to update the model. Some of the iterative schemes are

- Gradient descent method
- Steepest descent method
- Newton's method
- Conjugate gradient method

In general, we use the line search procedure for finding the optimal step length at three discrete points along the polynomial where it is at the minimum. Among all the methods, the steepest descent method allows slow convergence toward the optimum model and only takes first derivative information. Newton's method takes advantage of the second derivative to reach the optimum solution quicker than the steepest descent method. Due to fast convergence nature, the Hessian has to calculate at each iterative step. There are also different variants like damped and regularized in Newton's method. The practical advantage of the Newton's method over steepest descent method can be found in Pratt et al. (1998). Mainly, we terminate the iterative inversion based on stopping criteria such as the acceptable minimum error and/or imposing the maximum number of iterations. All of these methods suffer with local optimum solution and one has to start with very good starting assumption, which doesn't suffer with the existence of local solutions (Figure 4.2). Once we define the suitable objective function for the inverse problem, we always aim to find the global minimum for it. These methods are always trapped with local minima in cases of the starting model far away from the global minimum; thus, these methods are called local optimization methods.

We may need to adopt special inversion strategies like regularization and multi-scale approaches to avoid the local solutions. The regularization helps to implement the prior information and supports the need to develop the best possible method. In multi-scale approaches, we move toward global solutions by iteratively developing models in such a way that the previous models should satisfy the structural features of that particular wavelength and frequency order. All of the gradient methods require calculating the gradient of the objective functional with respect to model parameters in the vicinity of convexity properties (Figure 4.3).

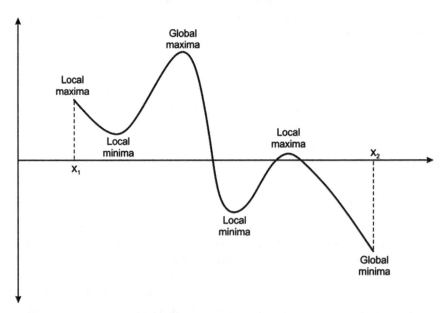

Figure 4.2 Differentiation of local and global solutions. Every global minima/maxima is a minimum/maximum of the polynomial but not vice-versa.

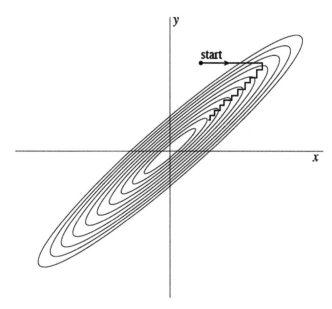

Figure 4.3 An example of gradient direction for the objective function converges toward the global minimum starting from the initial model.

On the other hand, global optimization techniques provide the solution in a global perspective, but it follows many optimum conditions depends on the scheme. Some of the global optimum methods are as

- Monte Carlo methods
- Simulated annealing methods
- Genetic algorithms
- Stochastic optimization methods

In each category of the above, we have many algorithms, and they follow their algorithmic steps. Extensive details of these methods can be found in Sen and Stoffa (2013).

5

Subsurface Parameterization

5.1 Introduction

The simulation of seismic wave propagation in a forward problem is done by assuming the material property through which it propagates in terms of numerical values. As in, defining the material property within the object (Earth) to model the seismic traveltime or wavefield from the source to the receiver is the so-called subsurface parameterization. Mostly in seismics, we use the velocity or slowness as a property that relates to traveltime. The subsurface is parameterized by many types of grids like regular grids and irregular grids (Figure 3.3) include node, blocks, and interfaces, etc. In block parametrization, we use the constant velocity for each block and in grid parametrization (Aki and Lee 1976; White 1989), we define the velocities at grid vertices along with interpolation between the grids (Thurber 1983). The block parametrization is always cautious when strong ray curvature is expected; on the other hand, we use this to define the higher order interpolation functions to overcome this difficulty for ray tracing. Rather than parametrizing the subsurface in the spatial domain, we can also define the spectral parameterization in the wave number domain in some form of truncated Fourier series (Wang and Pratt 1997). This parametrization is defined as infinitely differentiable velocity field by controlling the choice of the harmonic terms. The slowness distribution from the truncated Fourier series within a layer is represented as

$$s(r) = a_{00} + \sum_{m=1}^{N} \left[a_{m0} \cos(k.r) + b_{m0} \sin(k.r) \right] + \sum_{m=-N}^{N} \sum_{n=1}^{N} a_{mn} \cos(k.r) + b_{mn} \sin(k.r)$$

where r and k are location and wavenumber vectors respectively, and a_{mn}, b_{mn} are the amplitude coefficients of the (m, n)th harmonic term.

Sometimes, we can also parameterize the subsurface by incorporating the interfaces (for example, see section 5.3). This also helps to define geological structures

Active Seismic Tomography: Theory and Applications, First Edition. Kalachand Sain and Damodara Nara.
© 2023 John Wiley & Sons, Inc. Published 2023 by John Wiley & Sons, Inc.

into a model as prior information; it is most suitable for both reflection and refraction regimes (Williamson 1990; Zelt and Smith 1992). Irrespective of the subsurface parametrization, the user always tries to develop the model that explains the data's geological features. The subsequent sections explain one feature among them to build initial model.

5.2 Building Initial Model

We introduce pseudo workflow for a scheme where finite-difference is involved specially to incorporate prior information. The forward problem is solved using the finite-difference method by parameterizing the subsurface with uniform grids. A 1D velocity-depth function (Figure 5.1a) is enough for an initial model for seismic traveltime tomography where we expect long-wavelength velocity models similar to passive seismic tomography; however, this 1D velocity model may not be sufficient in areas where we expect finer details in complex terrains such as the thrust fold-belt and in volcanic rock covered regions and in complex sedimentary basins. Further, it is not possible to incorporate prior information to mitigate non-linearity in seismic tomography. Therefore, it necessitates an accumulated workflow to construct appropriate initial models for delineating structural details of the subsurface (Nara and Sain 2018). Here, we use the well-known intercept-time method (Dobrin and Savit 1988) and a subsurface parameterization described by Zelt and Smith (1992) to construct a pseudo-2D model (Figure 5.1b) rather than a simple laterally homogeneous model (Figure 5.1a) for seismic tomography. This also facilitates avoiding the exceptions made in deriving the initial model by distributing 1D velocity models over the seismic horizons obtained from the available seismic sections of multi-channel seismic (MCS) data (Sibuet et al. 2016). First, we briefly review the intercept-time method and the Zelt and Smith (1992) parameterization before moving to the workflow to construct a better initial model. This workflow may also be extended to the 3D data (Figure 5.1c).

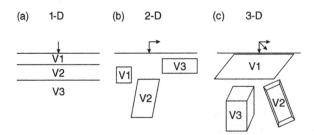

Figure 5.1 Representative models in different dimensions.

5.3 Intercept-time method

The intercept-time method provides the time-distance relation of the field seismic data. Figure 5.2a displays the time-distance plot for seismic head waves (Dobrin and Savit 1988).

The total time for the refraction path PQRS is $T_{PQRS} = \dfrac{x}{v_1} + \dfrac{2z\sqrt{v_1^2 - v_0^2}}{v_1 v_0}$; This represents a straight line on the time versus distance plot with a slope $\dfrac{1}{v_1}$ and the intercept-time $T_{inc} = \dfrac{2z\sqrt{v_1^2 - v_0^2}}{v_1 v_0}$ at x = 0. Note that, at the crossover distance of x$_{cross}$, the traveltimes for both layers are equal. Hence, we have the relation for interface calculating depth as $z = \dfrac{1}{2}\sqrt{\dfrac{v_1 - v_0}{v_1 + v_0}}\,x_{cross}$. Similarly, the generalized relation for multilayer and for different types of dipping interfaces can be obtained from Dobrin and Savit (1988).

5.4 Model Parameterization

For a trapezoid block, which has boundaries defined by $x = x_1$, $x = x_2$, $z = s_1x + b_1$, and $z = s_2x + b_2$ with the corner velocities v_1, v_2, v_3, and v_4 (Figure 5.2b), the velocity v within the trapezoid block can be expressed as

$$v(x,z) = \frac{\left(C_1 x + C_2 x^2 + C_3 z + C_4 xz + C_5\right)}{\left(C_6 x + C_7\right)}$$

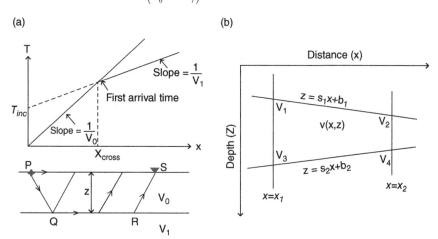

Figure 5.2 Representative (a) Illustration of time-distance plot and raypaths for head wave in a 2-layer case, and (b) Trapezoid block framed by juxtaposing the 1D velocity-depth functions.

where the coefficients C_i can be written as a linear combination of corner velocities.

$$C_1 = S_2(x_2v_1 - x_1v_2) + b_2(v_2 - v_1) - S_1(x_2v_3 - x_1v_4) - b_1(v_4 - v_3)$$

$$C_2 = S_2(v_2 - v_1) - S_1(v_4 - v_3)$$

$$C_3 = x_1v_2 - x_2v_1 + x_2v_3 - x_1v_4$$

$$C_4 = v_1 - v_2 + v_4 - v_3$$

$$C_5 = b_2(x_2v_1 - x_1v_2) - b_1(x_2v_3 - x_1v_4)$$

$$C_6 = (S_2 - S_1)(x_2 - x_1)$$

$$C_7 = (b_2 - b_1)(x_2 - x_1)$$

One can avail the advantages of using this parameterization, and more details can be obtained from Zelt and Smith (1992).

5.5 Workflow to Build Initial Model

- Pick the refraction/first arrival times, and compartmentalize the data
- Check the reciprocity and then plot the time-distance data
- Build the 1D velocity-depth models using the intercept-time method for each compartment
- Juxtapose all 1D models to construct pseudo-2D model
- Parameterize the model using the trapezoidal approach of Zelt and Smith (1992)
- Apply smoothing filter to remove the velocity discontinuities at the interfaces
- Grid the model uniformly for finite-difference modeling

This pseudo workflow used to define the better initial model along the geological features that are directly evident from the recorded seismic data without any bias. Thus, this helps to improve the mitigation of the local solutions and cycle-skipping problem, which challenges the modeler.

5.5.1 Example

Based on the proposed workflow, we perform first arrival traveltime tomography with the following three initial models to collect ocean bottom seismometer (OBS) data from the Indian offshore. The intercept-time method provides 1D sharp velocity models, which are smoothed by applying simple, moving average

Model-1: 1D model from all 65 OBS together (left panel) used in intercept-time method

Model-2: 1D model from all 65 OBS together (left panel) used in intercept-time method but model is distributed over the horizons picked from coincident seismic section

Model-3: Pseudo-2D model by juxtaposing 1D models (bottom panel) for different bunch of OBSs used in intercept-time method

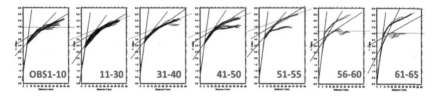

Figure 5.3 The 1D models construction from intercept-time method.

filters to allow seismic rays to pass a deep enough maximum through the model. The time-distance plots are shown in Figure 5.3. For Model-1 and Model-2, the first arrival data are picked from all 65 OBS gathers and are plotted together to build the possible 1D model. Whereas for Model-3, the first arrivals are selected from different compartments along the seismic line with aims to construct pseudo-2D velocity models.

Model-1: 1D velocity model, derived from all OBS gathers using the intercept-time method (Ex: Zelt and Barton 1998; Zelt et al. 2003)

Model-2: 1D velocity-depth function, derived from all OBS gathers using the intercept-time method, distributed over the horizons picked from the coincident seismic stack section (Ex: Sibuet et al. 2016)

Model-3: Pseudo-2D velocity model by juxtaposing various 1D velocity-depth functions derived from different compartments along the line.

By following the pseudo workflow, the developed initial models are shown in Figure 5.4. Model-1 shows only the homogeneous structural variations in the simplest variant of the model. Thus, it may easily trap local solutions compared to other models. Since, Model-2 was developed by utilising the reflection section interfaces, it will guide the convergence toward the global minimum better than Model-1. The availability of the reflection section may not always be possible, and it is also cost efficient in data acquisition. But, Model-3 is completely developed from the picked traveltime data alone, segment-wise; it clearly represents the geological structures on its own. This helps overcome all of the difficulties facing previous models and leads to a global minimum without becoming trapped at

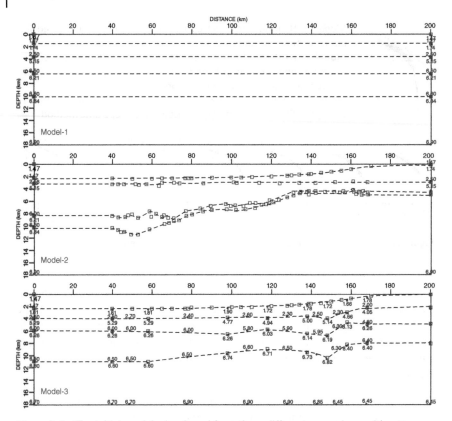

Figure 5.4 The initial models developed from three different ways along with our pseudo workflow are represented in Model-3.

local solutions. For this, the user is always required to observe the number of segments that are showing horizontal variations to develop the 2-D initial model for the inversion process.

The convergence history for traveltime tomography for all three models are shown in Figure 5.5. The convergence history of Model-2 and Model-3 shows that they are better fits, and they smoothly update the models with iterations compared to Model-1. We may not always have the coincident wide-angle OBS and multi-channel seismic (MCS) data to construct Model-2 as in the initial model; however, building Model-3 as an initial model from wide-angle OBS data is always possible.

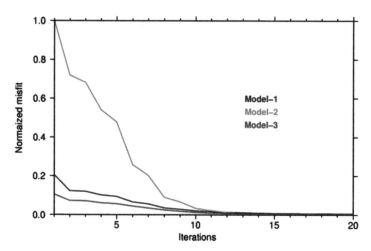

Figure 5.5 Convergence history for different initial models is used in traveltime tomography.

6

Seismic Travel Time Tomography

6.1 Introduction

The ray-based conventional forward modeling and inversion approaches to refraction data divide the entire crust into a few velocity layers. The study in which the model is parameterized by a selected number of velocity and boundary nodes may not provide the small-scale upper crustal velocity structure. Even so, it is difficult to derive a proper velocity model in a complex tectonic terrain, such as in the fold-thrust belt region or in basalt flow covered areas. The high amplitude source generated noise in deep crustal near-vertical reflections; profiling completely masks the shallow reflections and thereby leaves a gap in the uppermost shallow velocity structure. Therefore, it becomes difficult to correlate the deep crustal structure with the exposed geological features through this window, which is the link between the surface geology and the deeper subsurface structure. It is essential to delineate the finer details of the upper crust to understand the geo-tectonics and to identify the spatial locations of mineral resources in a region. The ray-based modeling techniques can not only delineate the finer structures, but they are limited in situations with a large volume of data like densely spaced marine data or 3D seismic refraction data. This necessitates using an alternate approach based on numerical modeling, which is not only fast and automatic but also can measure the resolution and uncertainty of estimated model parameters. Besides, modeling by numerical techniques can be used to derive velocity structure in any complex region. The cell-based tomography, which is borrowed from Medical Science, plays an important role in linking the surface geology with the subsurface structure.

Active Seismic Tomography: Theory and Applications, First Edition. Kalachand Sain and Damodara Nara.
© 2023 John Wiley & Sons, Inc. Published 2023 by John Wiley & Sons, Inc.

6.2   Traveltime Tomography

Traveltime tomography determines the seismic velocity structure on all scales, starting from the surface to the mantle; thereby, it has a wide range of applications from engineering problems to deep crustal and mantle studies. Seismic tomography is an inverse problem that involves searching the model space through which the rays are traced to minimize the misfit between observed and predicted traveltimes. However, the distinction between tomography and inversion is not clear; we thus use the uniform or grid parametrization to derive a smooth model from the seismic traveltime data since the P-waves are easy to pick from seismic data and used for deriving the subsurface velocity (Figure 6.1). Theoretically, both P- and S- wave traveltime tomography utilize same ray theory in forward modeling: thus, we can estimate the Poisson's ratio if both data are available for the same study.

We explain the tomographic approach (Zelt and Barton 1998; Zelt et al. 2003) to the first arrival seismic data along the profiles in brief. The First Arrival Seismic Tomography (FAST) approach was implemented by Zelt (1998a) and deals with both the forward and the inverse problems associated with the traveltime tomography. Basically, there are four steps to produce a tomographic image from seismic data.

   i) *Model parameterization:* Tomographic imaging is based on model parameterization with a set of unknown model parameters. It requires an initial model, which is parameterized by constant velocity blocks using a grid of velocity nodes.

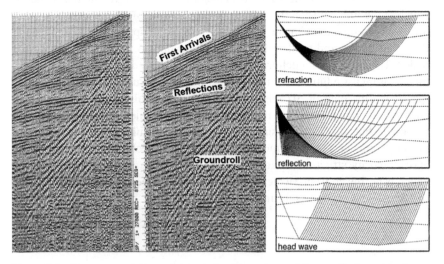

Figure 6.1   The representative seismic shot gathers along with raypaths traced for layer parameterization (Zelt 1999).

ii) ***Forward calculation***: A ray tracing technique suited to the forward step of the inverse approach is adopted for calculating model data, namely traveltimes, using a set of values for the model parameters and finding the source-receiver raypaths. The traveltimes and their partial derivatives with respect to velocity are calculated during raytracing using an efficient finite-difference numerical solution of raytracing eikonal equations (Hole and Zelt 1995; Vidale 1990).

iii) ***Inversion***: The inversion procedure automatically updates the model parameters with the objective of better fitting the model data to the observed data that is subjected to an adopted regularization procedure. The inverse step of adjusting model parameters is solved with the back projection method (Hole 1992; Zelt and Barton 1998). The traveltimes and raypaths are recalculated iteratively through the updated model until a satisfactory fit between the observed and modeled data, corresponding to a normalized chi-squared ($\chi2$) value of nearly one, is achieved.

iv) ***Analysis of solution robustness***: Analysis of the final model to know the robustness is carried out on covariance and resolution estimates. The final model is assessed for resolution and uncertainty/covariance with a checkerboard test using linear theory or on the reconstruction of test models that use same source-receiver field geometry using synthetic datasets.

6.3 History of Algorithms

One of the very first traveltime tomography algorithms for first arrivals is provided by Firbas (1981), and by White (1989) for regular grid parameterization, and by Spence et al. (1985) for later arrivals by irregular grids. Lutter et al. (1990) and Lutter and Nowack (1990) developed the algorithm to invert both first arrivals and reflections for velocity and interface independently with regular grid parameterization. Opposite to regular parameterization, a popular algorithm developed by Zelt and Smith (1992) facilitates the inversion of any type of the arrivals. Further Zelt and Smith (1992) allow for the incorporation of prior information with its irregular parameterization. Hammer et al. (1994) and Toomey et al. (1994) developed a 3D first arrival tomography algorithm to obtain a smooth velocity. Van Avendonk et al. (1998) and Korenaga et al. (2000) included the reflections, too, and with sheared grids, but these were limited to 2D applications. Zelt and Barton (1998) developed an algorithm for 3D first arrival tomography using smoothing regularization. The simultaneous refraction and reflection tomography algorithms developed by McCaughey and Singh (1997) and Hobro et al. (2003) solve for smooth velocities and interfaces with the allowance for discontinuities across the layer boundaries in both 2D and 3D applications (Figure 6.2).

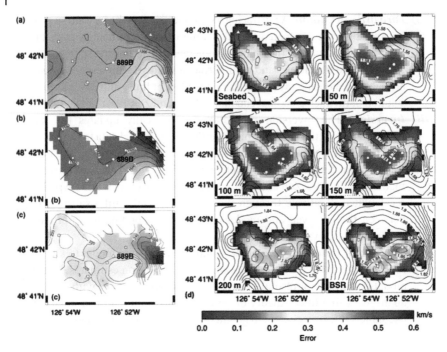

Figure 6.2 Representative 3D traveltime tomography results from both refraction and reflection data and coincident reflection profiles at Vancouver Island (Hobro et al. 2005). (a) Seafloor depth. (b) Bottom simulating reflector (BSR) depth. (c) BSR depth below the seafloor. (d) Slices through the final velocity model at the seafloor. Labeled contours indicate velocities (km/s) and color scale marks formal uncertainty estimates.

6.4 Objective Function

One of the most common forms used in traveltime tomography is regularization. This drives the iterative solution toward desirable data characteristics. One of the successful 2D regularized tomographic approaches was of Zelt and Barton 1998 and Zelt et al. 2003 to the first arrival seismic refraction data used to derive a minimum structure smooth velocity model. In general, the stability of inversion is enhanced by the number of model constraints. Regularization is an approach by which additional model constraints are applied to the inverse problem to treat the undetermined part of the solution (Scales et al. 1990). In regularized inversion, an objective function that measures a combination of data misfit and model roughness is minimized in the least-square sense (Menke 1989) to provide the smoothest model. The objective function $\phi(s)$, minimized at each iteration, is given by

$$\phi(s) = \delta t^T C_d^{-1} \delta t + \lambda \left[\alpha \left(s^T W_h^T W_h s + s_z s^T W_v^T W_v s \right) + (1 - \alpha) \delta s^T W_p^T W_p \delta s \right] \quad (6.3)$$

where δt is the traveltime misfit vector, s is the model slowness vector, and $\delta s = s - s_0$ is the perturbed model vector with s_0 being the staring slowness, C_d being the data covariance matrix that contains the estimated pick uncertainties, W_p is the perturbation weighting matrix, and W_h and W_v are the horizontal and vertical roughness matrices, respectively. The perturbation weighting matrix is a diagonal matrix of the reciprocal starting slowness values, measuring the relative perturbation of the current model from the starting model. The roughness matrices are second-order spatial finite-difference operators that measure the roughness of the model in the horizontal and vertical directions. The parameters λ, α and s_z control the weight of the terms in the objective function. λ is the trade-off parameter that controls the overall regularization and is reduced systematically, starting from λ_0, using a reduction factor λ_r. It constrains the long wavelength model structure in initial iterations and the fine-structure in subsequent iterations. The s_z determines the relative weight of the vertical and horizontal smoothing regularization. The parameter α determines the relative weight of the second derivative and perturbation regularization of velocity parameters. Minimizing the objective function with respect to the unknown model parameters (v and z) leads to a linear system of equations that are solved using the least-square variant of conjugate gradient algorithm (Nolet 1987).

A laterally homogeneous 1D velocity model is used as a starting model for tomographic inversion and the model and ray paths are upgraded over several iterations. At each iteration, first arrival travel-times are calculated at each receiver location using the finite-difference solution to the eikonal equation based on the approach of Hole and Zelt (1995) to handle large velocity contrasts. After traveltime computation, ray paths are determined by following the traveltime gradient from a receiver back to the source (Vidale 1988). At each iteration, the objective function $\varphi(s)$ (Equation 6.3) is minimized in a least-square sense. The iteration continues until the normalized $\chi 2$ misfit between the observed and predicted traveltimes reduce to a value of almost one. The normalized $\chi 2$ misfit is given by

$$\chi^2 = \frac{1}{N} \sum_{i=1}^{N} \left[\frac{t_i^o - t_i^p}{\sigma_i} \right]^2$$

where N is the number of traveltimes used, t_i^o and t_i^p are the ith observed and predicted traveltime, and σ_i is the assigned picking uncertainty. The entire procedure is explained by a flow chart in Figure 6.2.

6.5   Model Assessment

Once the model is obtained by any algorithm, we need to assess it to glean how well we delineated the resolution, error and non-uniqueness, etc. We assess the model in terms of ray coverage and of ray hit counts using ray plots. The spatial resolution or single model parameter of the derived model can be assessed by inverting the synthetic data for the final model and the observed data with the same parameter set including the trade-offs model parameters (Zelt 1999; Zelt and White 1995). This technique allows us to observe that the amount the adjacent parameters differs from their original positions in the final model (Figure 6.3) and remember that the picking uncertainty yields a significant traveltime anomaly. Unlike the single parameter resolution checking, we can also assess the spatial resolution of the entire model with the test popularly known as checkerboard test (Zelt 1998b; Zelt and White 1995). We can find the numerous modeling strategies, and direct and indirect model assessment techniques from Zelt (1999) and Rawlinson and Sambridge (2003).

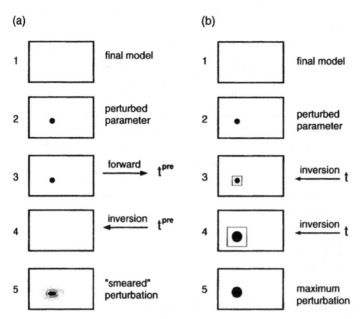

Figure 6.3   Spatial resolution of the model through the single model parameter resolution and uncertainty test in five major steps (Zelt 1999). Step 1: Final derived model. Step: Selection of the intended model parameter (Black dot). Step 3, 4 and 5 in (a) are to estimate the spatial resolution of the chosen model parameter. In (b) are to estimate the uncertainty of the chosen model parameter.

7

Seismic Full Waveform Tomography

7.1 Introduction

In exploration seismology, the traveltimes were earlier determined based on ray-tracing (Cerveny 1987, 2001) till the introduction of wavefront tracking schemes through the normal/staggered grid parameterization (Qin et al. 1992 and Vidale 1990). This included the finite-difference solutions of the Eikonal equation and thus it became one of the most powerful geophysical techniques for providing high resolution subsurface images in order of Fresnel zone. Several authors started the usage of whole data by including all arrivals, which produced higher resolution images of the order of sub-wavelength. This was named full waveform tomography (FWT).

As the crustal velocity structure is exploited using the traveltime tomography of selected phases from the whole seismograms, the small-scale structures or features cannot be resolved. It is to be stated that the resolution of the traveltime tomography is in the order of Fresnel zone. Whereas, all information (seismic velocity, density, attenuation, and anisotropy) is contained into the traveltimes, amplitudes, frequencies, and phases of the whole seismograms, which are exploited by FWT wherein the resolution of the order of sub-wavelength is achieved. The wavelength is much smaller than the Fresnel zone.

A lot of theoretical works on FWT have been carried out over the past 20 years; still it remains elusive as the application mostly lies in the theoretical domain because of the complexity involved in understating and implementing the physics of the waves. There exists a tremendous scope for applying FWT to the field data. It computes the difference between the observed and the synthetic response of the current velocity model and of the processes in a way that is similar to reverse time migration (RTM) of the residual dataset to compute a gradient volume (Lailly 1983 and Tarantola 1984) and to update the velocity model. The gradient of the misfit function is computed by correlating the forward-propagated wavefields emitted

Active Seismic Tomography: Theory and Applications, First Edition. Kalachand Sain and Damodara Nara.

from the source with the backward propagated data residual wavefields emitted from the receivers without explicitly computing the partial derivatives. Numerical results from this approach are given by Kolb et al. (1986), Gauthier et al. (1986), and Mora (1987) in the time domain. Shin (1988), Pratt and Worthington (1990), Pratt (1990a, 1990b), Geller and Hara (1993), and Song et al. (1995) apply the same idea to inverse problems in the frequency-domain by using implicit frequency-domain numerical algorithms.

An excellent technical review of FWT is given by Virieux and Operto (2009). A gradient based, iterative algorithm for FWT is available in earlier works by Lailly (1983), Tarantola (1984), Worthington (1984), and Mora (1987). There are many reviewed works on seismic tomography that include several strategies for modeling and inverting both traveltimes and full waveforms. Here, we mention some of those: Nowack and Braile (1993), Pratt (1999), Pratt and Shipp (1999), Zelt (1999), Carcione et al. (2002), Rawlinson and Sambridge (2003), Lavander et al. (2007), and Virieux and Operto (2009).

7.2 Full Waveform Tomography (FWT)

The FWT uses a two-way wave equation to produce high resolution velocity models. The key aspect of this method is that the wavefields recorded at the surface must be accurately modeled to represent the kinematics, and, to a limited extent, the dynamics of all the waves during the iterative inversion toward the final Earth model.

We intend to apply the full waveform tomography or inversion in the frequency-domain using the classic gradient method (Pratt et al. 1998) in which the full-wave equation is used rather than a one-way wave equation that ignores backscattering and wide-angle effects. From the spatial discretization, the acoustic or elastic wave equations using finite-difference or finite-element approach turns into an algebra-differential equation in the time-space domain (Murfurt 1984) that can be written as

$$\mathbf{M}.\ddot{\mathbf{u}}(\mathbf{t}) + \mathbf{K}.\bar{\mathbf{u}}(\mathbf{t}) = \bar{\mathbf{f}}(\mathbf{t}) \tag{7.1}$$

M and K are referred to as the mass matrix and the stiffness matrix respectively. $\bar{\mathbf{u}}(\mathbf{t})$ is the discretized wavefield (i.e., pressure or displacement) as a column vector, and $\bar{\mathbf{f}}(\mathbf{t})$ represents a discrete version of the force density or the source as a column vector. Using the temporal Fourier transformation of the full wave equation to implement in frequency-domain (Pratt 1990a and Pratt et al. 1998), the resultant numerical system can be represented briefly by

$$-\omega^2\mathbf{M}.\bar{\mathbf{u}}(\omega) + \mathbf{K}.\bar{\mathbf{u}}(\omega) = \bar{\mathbf{f}}(\omega) \tag{7.2}$$

This can be simply written in linear form as

$$\mathbf{S}(\omega)\,\mathbf{u}(\omega) = \mathbf{f}(\omega) \text{ or } \mathbf{u}(\omega) = \mathbf{S}^{-1}(\omega)\,\mathbf{f}(\omega) \tag{7.3}$$

Where u is the Fourier-transformed, complex-valued, and a discretized wavefield (the pressure or displacement vector); $\mathbf{S}$ is a complex-valued impedance matrix and is given as $\mathbf{S} = \mathbf{K} - \omega^2\mathbf{M}$ where f is the source term.

The weighted least-squares linearized waveform inversion is solved in the frequency-domain by a classic gradient method (Pratt et al. 1998). The updated model $\boldsymbol{m}$ is obtained by the relation $m = m_0 + \Delta m$ where m_0 is the starting model, and Δm is the model perturbation obtained from the weighted least-squares, with some scaling and some with regularization to the gradient for reliable perturbation models. This is given by

$$\Delta m_i = -\alpha(diagH_a + \epsilon I)^{-1} g_m Re\left\{ u^t \left[\frac{\partial S^t}{\partial m_i} \right] S^{-1} W_d \Delta d^* \right\} \tag{7.4}$$

where H_a is the Hessian matrix, g_m is the spatial smoothing operator, and W_d is the weighting operator applied to the data. S is the green function for the source located at each grid node of the velocity model, and α is the step length that controls the amplitudes of the perturbations. Equation (7.4) gives the model perturbation for one frequency and one shot gather. One can obtain the expressions for multiple shot gathers with simple summation over shots for the diagonal Hessian and gradient.

$$\Delta m_i = -\alpha \sum\nolimits_{ns} (diagH_a + \epsilon I)^{-1} g_m \sum\nolimits_{ns} Re\left\{ u^t \left[\frac{\partial S^t}{\partial m_i} \right] S^{-1} W_d \Delta d^* \right\} \tag{7.5}$$

Similarly, for multiple frequencies, the expression can be written by summation over the frequencies.

$$\begin{aligned}\Delta m_i = -\alpha \sum\nolimits_{nw} \sum\nolimits_{ns} (diagH_a + \epsilon I)^{-1} \sum\nolimits_{nw} gm \\ \sum\nolimits_{ns} Re\left\{ u^t \left[\frac{\partial S^t}{\partial m_i} \right] S^{-1} W_d \Delta d^* \right\}\end{aligned} \tag{7.6}$$

For more details on the inverse problem, one can refer Pratt et al. (1998), Pratt (1999), and Operto et al. (2006).

7.3 Frequency-Domain FWT Algorithm

The algorithmic steps implemented in FWT are briefly explained in following three stages:

Stage 1: Initial model building Since the FWT has a variety of difficulties such as computational costs, non-linearity of the problem, and the inherent use of the Born approximation in finding the gradient direction and slow convergence rate, etc., it needs a good starting model. Furthermore, the FWT develops the structures at or below the dominant seismic wavelength from the long wavelength subsurface structures. In general, the long wavelength velocity models are constructed from traveltime tomography.

Stage 2: Pre-processing To implement the computationally efficient FWT, data-preprocessing is necessary to meet the approximations used in both the forward and inversion algorithms of FWT. The approximations include acoustic wave equation, constant density, and quality/attenuation factor Q to be used due to the highly sensitive nature of multi-parameter full waveform tomography. The pioneers of the FWT have designed the following investigative steps because of difficulties in accounting for the acoustic approximation (Operto et al. 2006 and Ravaut et al. 2004).

- Trace editing to avoid noisy traces for quality control
- Minimum-phase frequency-domain deconvolution and spectral amplitude normalization to its maximum (Yilmaz 2001)
- Band-pass filtering for improving the signal-to-noise (S/N) ratio
- Strengthen the lateral trace coherency and improve the S/N ratio using the coherency filtering using spectral matrix filtering
- Time windowing to eliminate PS-converted waves and deep reflections correspondence to outside the bounds of model
- Offset windowing to remove ground roll in dealing with land seismic data and to exclude traces where the water wave is dominant in marine seismic data from short-offset range
- Converting 3D geometrical spreading into 2D amplitude behavior by multiplying with $\sqrt{t}$

Stage 3: FWT algorithm

- Start loop over group of frequencies
  - Start loop over imaginary part of frequencies
    - ○ Start loop over iterations
      Read starting model
      - ■ Start loop over frequencies in one group
        Build Impedance matrix
        Part 1: Compute Diagonal Hessian
        Part 2: Compute Gradient
      - ■ End loop over frequencies in one group
        Scale the Gradient

Part 3: Compute step length
Update the model
  ○ End loop over iterations
  – End loop over imaginary part of frequencies
• End loop over group of frequencies

The entire FWT algorithm is represented by a flow chart as given in Figure 7.1. The FWT algorithm is written in FORTRAN90 and is parallelized with the Message Passing Interface (MPI) by the domain-decomposition approach. The system of equations is solved by using the massively parallel direct solver MUMPS.

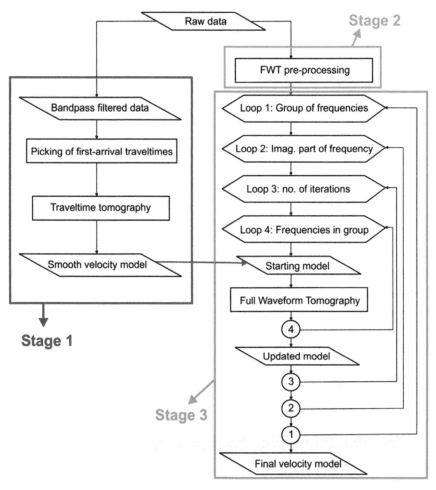

Figure 7.1 Flow chart of full waveform tomography. (modified after, Operto et al. 2004).

7.4 Frequency Selection Criteria

Generally, to reconstruct the images, we need to recover the wavenumber coverage of the source-receivers. That can be done either by using a wide band of frequencies or by using different source-receiver pairs to sample the same imaging position in different directions since the temporal bandwidth of the acquisition is directly related to the wavenumber bandwidth of the imaging point. Thus, we can optimally choose the selective set of frequencies, which can not only avoid the redundancy of the wavenumbers but can reduce the computational resources as well. Let us consider a monochromatic incident and scattered plane waves propagating in the direction $\hat{s}$ and $\hat{r}$ with the homogeneous background model of velocity c_0. The corresponding single wavenumber vector component is represented as $k = k_0(\hat{s} + \hat{r})$ and follows the symmetric angles for both incident and scattering waves (Figure 7.2a). We can define these for the 1D case as

(a)

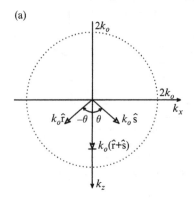

(b)

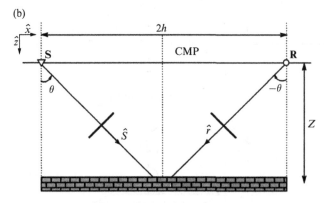

Figure 7.2 The simple configuration for wavenumber illumination (Sirgue and Pratt 2004).

$$k_0 \hat{s} = (k_0 \sin\theta, \ k_0 \cos\theta) \ and \ k_0 \hat{r} = (-k_0 \sin\theta, \ k_0 \cos\theta) \tag{7.7}$$

From Figure 7.2b, we have

$$\sin\theta = \frac{h}{\sqrt{h^2 + z^2}} \ and \ \cos\theta = \frac{z}{\sqrt{h^2 + z^2}} \tag{7.8}$$

Now the wavenumber vector $k = k_0(\hat{s} + \hat{r})$ components become

$$k_x = 0 \ and \ k_z = 2k_0 \cos\theta = 2k_0 \alpha$$

$$\alpha = \cos\theta = \frac{z}{\sqrt{h^2 + z^2}} = \frac{1}{\sqrt{1 + R^2}} \tag{7.9}$$

where $R = \frac{h}{z}$ defines the half offset-to-depth ratio. It implies, for larger than the R value, that the vertical wavenumber contribution is smaller. Thus, the vertical wavenumber recovery will be more for a given frequency and offset range (Sirgue and Pratt 2004). The maximum and minimum vertical wavenumbers contributed from the nearest and the largest offsets. Thus, we have these in terms of frequency as

$$k_{zmax} = \frac{4\pi f}{c_0} \ and \ k_{zmin} = \frac{4\pi f \alpha_{min}}{c_0}$$

and the wavenumber bandwidth for multioffset acquisition,

$$\frac{k_{zmax}}{k_{zmin}} = \frac{1}{\alpha_{min}} = \sqrt{1 + R_{max}^2}$$

This implies that for the increasing frequency, the wavenumber coverage will increase (Sirgue and Pratt 2004 and Wu and Toksoz, 1987). Where, $R_{max} = \frac{h_{max}}{z}$, h_{max} defines the maximum half offset and z defines the depth to the target layer. Thus, in order to keep the wavenumber continuity and to select the optimal frequencies, we choose an iterative strategy as

$$k_{zmax}(f_n) = k_{zmin}(f_{n+1})$$

where f_n is the current frequency to choose the next frequency f_{n+1}. The minimum wavenumber coverage of the larger frequency f_{n+1} must be equal to the maximum wavenumber of the smaller frequency f_n; this can be observed in Figure 7.3.

$$f_{n+1} = \frac{k_{zmax}}{k_{zmin}} f_n = \frac{f_n}{\alpha_{min}}$$

From the Figure 7.3 relation, the frequency increment follows the relation

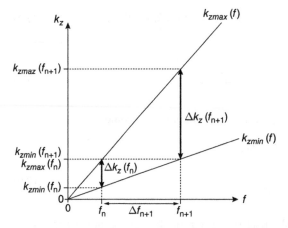

Figure 7.3 Illustration of frequency selection for FWT based on vertical wavenumber coverage (Sirgue and Pratt 2004).

$$\Delta f_{n+1} = f_{n+1} - f_n = \left(\frac{1}{\alpha_{min}} - 1\right) f_n = \left(\frac{1 - \alpha_{min}}{\alpha_{min}}\right) f_n = (1 - \alpha_{min}) f_{n+1}$$

The Figure 7.3 relation tells us that the increment between the subsequent frequencies are not constant and also that they increase with frequency. This frequency selection criteria helps to reduce the computational time efficiently, which has successfully been applied in multiscale imaging FWT (Brenders and Pratt 2007b).

Section II

Applications

8

Tomography of Synthetic Seismic Data

8.1 Introduction

In general, the researchers demonstrate the efficacy of FWT algorithms in the frequency-domain FWT of synthetic data by using preferential frequencies and smoothed version of the true model as the preferred starting model, which is very close to the solution. Implementation of the same algorithm and strategies for both generating and inverting the data and preferential use of frequencies in frequency-domain FWT of synthetic data is difficult to select for FWT of field seismic data (e.g., Ben-Hadj-Ali et al. 2008, Brossier et al. 2009, Ravaut et al. 2004, Shin and Cha 2008, Sourbier et al. 2009, and Vigh and Starr 2008). To understand the practical aspects during FWT, we would like to study the sensitivities of different factors such as the starting model, frequency selection, and their overlapping of multiscale imaging in the frequency-domain FWT. In this type of study, it is very important to know the intricacies involved in the FWT of field seismic data. In addition, one must keep in mind that the FWT is also sensitive to other factors that include the algorithm that has been implemented, the solution of the inverse problem, the cost function, the step-length criteria, etc. However, these factors are outside the scope of our study and need further attention in future research.

We have organized our study into four sub-headings. In section 8.2, we describe the true model, which was used to generate ocean bottom seismometer data by finite-difference elastic modeling. We then apply the acoustic full waveform tomography to the elastic synthetic data in section 8.3. The results of traveltime and FWT are discussed in section 8.4. The implications of the study are summarized in section 8.5. The outline of the study is described with a chart as shown in Figure 8.1.

Active Seismic Tomography: Theory and Applications, First Edition. Kalachand Sain and Damodara Nara.

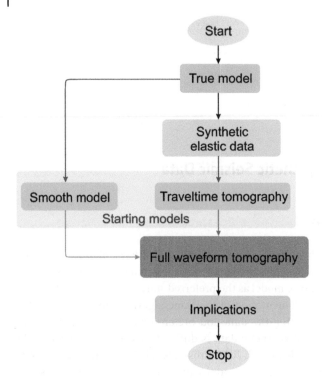

Figure 8.1 Flowchart of acoustic full waveform tomography from the elastic synthetic data.

8.2 Forward Modeling

We constructed a 2D P-wave velocity (V_p) model (Figure 8.2a) by correlating two available sonic logs in a marine environment. The true model shows finer details of the subsurface and is complex in nature. This can be considered one of the benchmark models for pursuing research in seismic tomography. The S-wave velocity (V_s) and density (ρ) models, derived from the V_p model using the following relations (1) and (2), are shown in Figures 8.2b and 8.2c, respectively.

$$V_s(m/s) = 0.8619 x V_p - 1172 \tag{8.1}$$

$$\rho(g/cc) = 0.31 x \left(V_p\right)^{0.25} \tag{8.2}$$

The dimensions of the model are 50.0 km in length and 5.0 km in depth. It is parameterized by 12500×1250 nodes for forward modeling using 4 m grid spacing, which may be sufficient to represent the true model; however, the model has very fine-scale structural details in the order of log-scale. One can notice that two very important dominant features such as (i) high velocity pinch-out and (ii) dominant convex features at depths of 3.0 km and 4.0 km respectively with intermediate hidden zones between them. Also we marked the major interfaces observed in the model (Figure 8.2). The wide-angle elastic data were generated using finite-difference modeling (Bohlen 1998, 2002) for this complex geological model. The

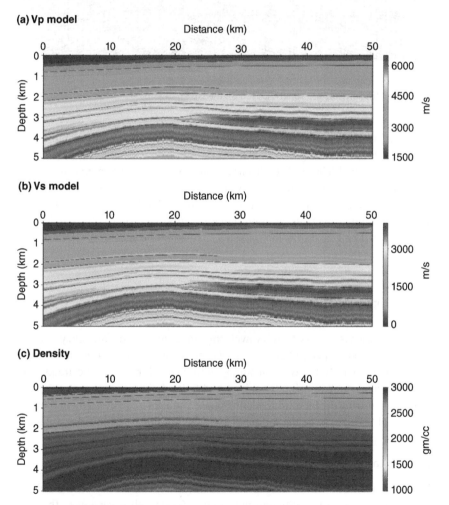

Figure 8.2 True (a) P-wave velocity (Vp), (b) S-wave velocity (Vs) and (c) Density models used for generating the elastic data. Identified major interfaces are shown by thin black lines.

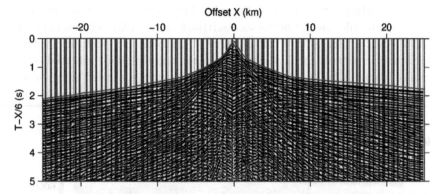

Figure 8.3 A representative OBS gather is displayed with a reduction velocity of 6.0 km/s. The picked first arrivals are superimposed on the OBS gather.

synthetic seismic experiment includes a total of 51 ocean bottom seismometers (OBSs) at 1.0 km spacing and 1000 shots (at 5 m below water surface) per OBS with 50 m interval. The OBS depth varies from a minimum of 45 m to a maximum of 370 m. The normalized Ricker wavelet with a central frequency of 10 Hz was used to generate the synthetic data. The total recording length is 16.384 *s* with a sampling interval of 4 *ms*. A representative OBS gathering with a reduction velocity of 6.0 km/s is shown in Figure 8.3. Hereafter, the generated elastic data is referred to as the realistic seismic data throughout the study.

8.3 Visco-Acoustic Full Waveform Tomography

We applied the visco-acoustic full waveform tomography in the frequency-domain (Dessa et al. 2004, Operto et al. 2004, 2006, and Ravaut et al. 2004) to the realistic seismic data. The full waveform tomography algorithm solves the frequency-domain wave equation using the mixed-grid finite-difference approach (Hustedt et al. 2004) during the forward modeling phase and uses a classic gradient method (Pratt et al. 1998) to solve the weighted least-squares linearized inverse problem.

8.3.1 Starting Model

For any gradient search inversion, we need a starting model. We constructed the starting model from two different approaches: (i) smoothed version of the true model and (ii) the first arrival traveltime tomography model. The first one is straightforward, people generally use it to demonstrate the efficacy of seismic tomography during the development of an inversion algorithm. For field experiments, the true velocity structure is never known, but many researchers use only the smooth version of the true model. This avoids the situation of building an initial model from synthetic data as in field experiments by any practical approaches (e.g. first arrival/reflection traveltime tomography and stereo-tomography). Meanwhile, the later one is considered for field application in which the subsurface model is unknown. We focused on both types of initial models in this study.

We employed the well-known first arrival seismic tomography (FAST) code (Zelt 1998a, and Zelt and Barton 1998) to the traveltime data picked from all OBSs. This package offers iterative regularized inversion in terms of data misfit and roughness and calculates new raypaths at each iteration. For more details, the reader may refer to recent works of Zelt et al. (2003), Vijaya Rao et al. (2015), and Damodara et al. (2017). A total of 50,903 first arrival picks with a 30 ms uncertainty are used. We parameterized the model in 201 × 21 grids having 0.25 × 0.25 km grid size in forward modeling and 0.5 × 0.25 km cell size during inversion. The final traveltime tomographic model (Figure 8.4) was achieved after 16 iterations with a root mean square traveltime residual of 33.8 ms from a starting model that had been derived with the intercept-time method by taking different OBS data and juxtaposing them (Damodara et al. 2017). The corresponding raypaths for the final model and traveltime residual plots for both the starting and final models are

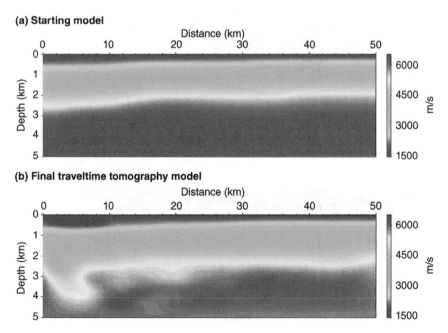

Figure 8.4 (a) Starting model and (b) Final velocity model from first arrival traveltime tomography.

shown in Figure 8.5. The ray paths plot (Figure 8.5a) clearly shows that we have good ray coverage and penetration up to depths of ~ 3.5–4.0 km and that we also don't have any ray penetration at corners of the profile. It indicates that the delineated model (Figure 8.4b) has good confidence at ray penetrated positions;

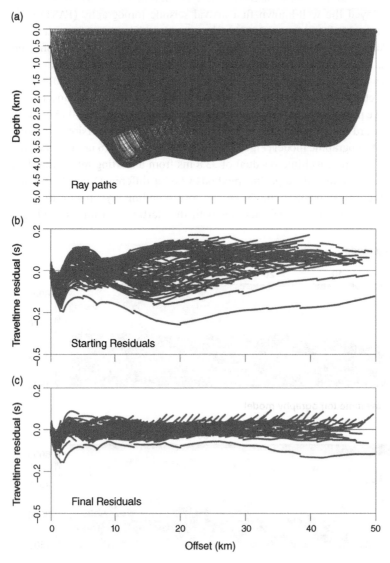

Figure 8.5 (a) Ray tracing through the final traveltime tomographic model. Traveltime residual for (b) Starting and (c) Final models respectively.

remaining positions of the model represent the background model only. This traveltime tomography (TT) model was used as one of the starting models for FWT, which allows that the inherent propagation error that may not require adding, which is explicitly similar to the Gaussian noise during synthetic data generation. The other starting model (Figure 8.6) is the smoothed version of the true model obtained by applying damped least-squares moving average filter on the true model. The dimensions of the filter 50 and 50 samples were used in both directions (Distance and Depth). Both the starting models (Figures 8.4b and 8.6) show the long wavelength features but lack in finer details; we expect these from full waveform tomography.

8.3.2 Data Pre-Processing

Since we apply the full waveform tomography in the acoustic sense, data pre-processing is necessary to mitigate or to exclude arrivals and propagation effects in data, which cannot be predicted from acoustic approximation. A time window of 1300 ms was applied around the first arrivals to exclude the non-acoustic arrivals such as the P-SV converted arrivals, free-surface multiples and early ambient noise, etc. A representative pre-processed OBS gather is displayed in Figure 8.7. The data were subsequently Fourier transformed, and selected components were extracted in order to form a mono-frequency dataset for full waveform tomography in the frequency-domain.

8.3.3 Full Waveform Tomography

Many of the past FWT case studies (Brenders and Pratt 2007b and Wang and Rao 2009) verified that a discrete set of a few frequencies are enough to obtain reasonable results from FWT in frequency-domain. Earlier studies followed two

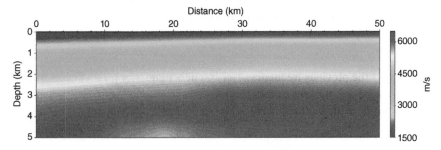

Figure 8.6 Smoothed version of the true model, achieved by applying damped least-squares moving average filter on the true model. The dimensions of the filter 50 and 50 samples were used in both directions (Distance and Depth).

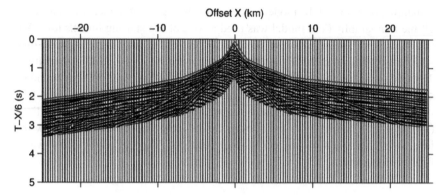

Figure 8.7 A representative pre-processed OBS gather displayed with a reduction velocity of 6.0 km/s. The picked first arrivals are superimposed on the OBS gather.

strategies: (i) a discrete set of frequencies with an equal interval (Dessa et al. 2004, Gorszczyk et al. 2017, Malinowski and Operto 2008, and Operto et al. 2006) and (ii) a discrete set of frequencies selected by the criteria of Sirgue and Pratt (2004). We briefly describe the later strategy in the next sub-section.

8.3.3.1 Frequency Selection Criteria
As the computational cost increases over a range of frequencies starting from low frequency to high frequency in full waveform tomography, Sirgue and Pratt (2004) demonstrated that a properly selected limited number of frequencies are enough for full waveform tomography by the utilization of wavenumber spectrum. Brenders and Pratt (2007b) termed the full waveform tomography using the selected frequencies (Sirgue and Pratt 2004) as an efficient waveform tomography. This strategy has been applied to numerous studies by Brenders and Pratt (2007b), and Wang and Rao (2009). We also adopted the similar strategy in our visco-acoustic full waveform tomography. To have continuous wave number coverage, the proposed strategy for frequency discretization is

$$f_{n+1} = \frac{f_n}{\alpha} \tag{8.3}$$

Where f_n is the previous frequency, f_{n+1} is a new frequency for inversion, and α is the cosine of the maximum incident angle for a scattered wavefield, corresponding with the maximum value of the half source-receiver offset h_{max} and target depth z.

$$\alpha = \frac{z}{\sqrt{h_{max}^2 + z^2}} \tag{8.4}$$

8.4 Results and Discussion

Since our focus is to study the influence of the starting model and frequency selection and their combinations in producing the best possible results from visco-acoustic FWT in the frequency-domain, we restricted use to all of the same other parameters except those mentioned above. Figure 8.8 summarizes where the cases differ from each other during the multiscale imaging. Further, we used constant density and Q for all case studies. The study by Kurzmann et al. (2013) shows that for single parameter acoustic-inversion, the constant models (density and Q other than the inverted parameter V_p) are enough to obtain better results. The case studies are described below.

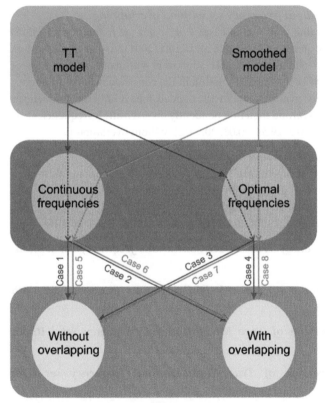

Case 1: TT model-Continuous freq. set without overlapping
Case 2: TT model-Continuous freq. set with overlapping
Case 3: TT model-optimal freq. set without overlapping
Case 4: TT model-optimal freq. set with overlapping
Case 5: Smooth model-Continuous freq. set without overlapping
Case 6: Smooth model-Continuous freq. set with overlapping
Case 7: Smooth model-optimal freq. set without overlapping
Case 8: Smooth model-optimal freq. set with overlapping

Figure 8.8 Flow chart for different strategies of FWT.

Case 1: TT Model-Continuous frequency set without overlapping

In this case, the traveltime tomography model is a starting model, and a continuous, discrete set of 40 frequencies starting from 2 Hz to 21.5 Hz with an interval 0.5 Hz were used. These 40 frequencies are formed into 20 groups (e.g., group 1: 2, 2.5 Hz; group 2: 3.0, 3.5 Hz; group 3: 4.0, 4.5 Hz, etc.) with each group having two frequencies without overlapping between the frequency groups.

Case 2: TT Model-Continuous frequency set with overlapping

We used the same starting model and discrete set of frequencies as in Case 1, but they were formed into 39 groups by overlapping between the frequency groups with one common frequency (e.g., group 1: 2, 2.5 Hz; group 2: 2.5, 3.0 Hz; group 3: 3.0, 3.5 Hz, etc.).

Case 3: TT Model-Optimal frequency set without overlapping

We used the same starting model as in cases 1 or 2, and the frequency selections were done using the criteria set by Sirgue and Pratt (2004). As per this criterion, only the two frequencies, 2.0 Hz and 10.2 Hz, that were within our interest range of between 2.0–20.0 Hz will contribute. To maintain the consistency in the frequency range and stabilize the inversion, we further added four additional frequencies (1.95, 19.54, 20.03, 20.1 Hz) and then independently inverted all six frequencies one after another (1.95, 2.0, 10.2, 19.54, 20.03, 20.1 Hz) without overlapping.

Case 4: TT Model-Optimal frequency set with overlapping

The same strategy was used as in case 3 except frequency overlapping. We formed all six frequencies into four groups with each group having three frequencies due to overlapping between the groups (e.g., group 1: 1.95, 2.0, 10.2 Hz; group 2: 2.0, 10.2, 19.54 Hz; group 3: 10.2, 19.54, 20.03 Hz; group 4: 19.54, 20.03, 20.1 Hz).

Case 5: Smooth version of True Model-Continuous frequency set without overlapping

Case 6: Smooth version of True Model-Continuous frequency set with overlapping

Case 7: Smooth version of True Model-Optimal frequency set without overlapping

Case 8: Smooth version of True Model-Optimal frequency set with overlapping

Case 5, Case 6, Case 7, and Case 8 are similar to Case 1, Case 2, Case 3, and Case 4, respectively, except the starting model, which is the smoothed version of the true model.

The final models, obtained from FWT for each case as described in Figure 8.8 are shown in Figure 8.9. The 1D velocity-depth functions extracted from FWT models at 17.9 km distance are compared with those extracted from the true model, generated from well log data, and are shown in Figure 8.10.

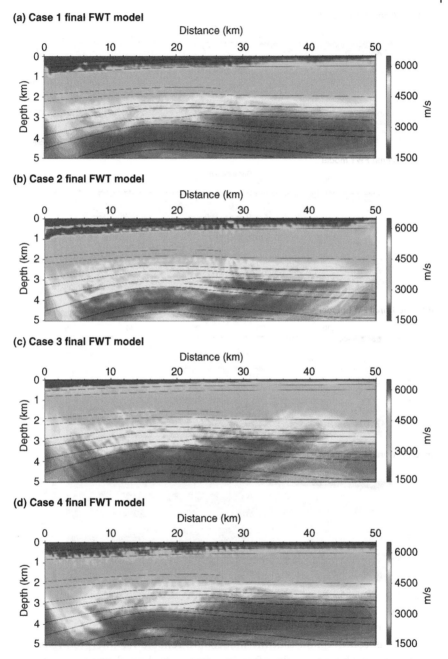

(a) Case 1 final FWT model

(b) Case 2 final FWT model

(c) Case 3 final FWT model

(d) Case 4 final FWT model

Figure 8.9 Final FWT models obtained from each FWT case as described in Figure 8.8. Major interfaces are superimposed on the final FWT models.

(e) Case 5 final FWT model

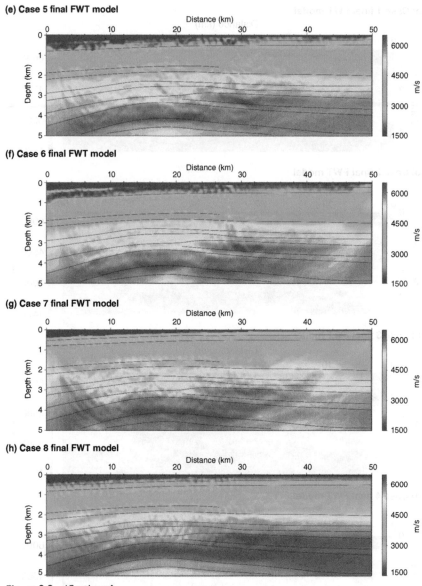

(f) Case 6 final FWT model

(g) Case 7 final FWT model

(h) Case 8 final FWT model

Figure 8.9 (*Continued*)

In general, everyone sees having a good correlation with the available well log data as confirmation that the models are reliable for interpreting/postulating further about the subsurface in terms of geological features. But, this may lead the researchers in the wrong direction as we will be observing in our 1D velocity-depth functions (Figure 8.10) at two well positions (17.9 km and

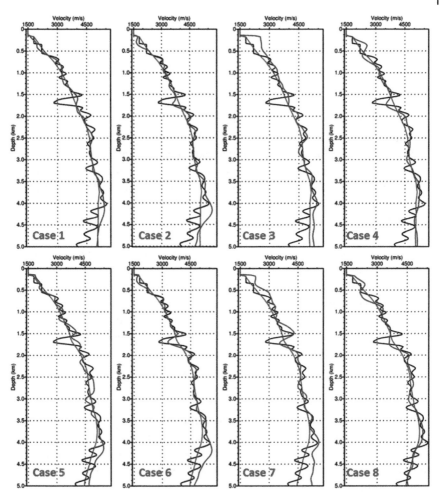

Figure 8.10 1D-models extracted from the true (black) traveltime tomography (red) and FWT models (blue) from each case at 17.9 km.

28.9 km) because all 1D velocity-depth functions of FWT are almost close enough (at least 60–70%) to true models at both well positions except at very few depth positions. Expecting the outstanding correlation with well log information are also not fair due to several factors such as the usage of band-limited data, discrete frequency inversion, difference between the log and seismic scale, and much more. Note that, we have performed FWT with consistent parameters such as both modeling and inversion for each FWT case, and it may terminate updating of the FWT model before the maximum number of iterations due to the deterministic stopping rule. Thus, it may also inherently effect the model update for same initial model, irrespectively.

All of the final 2D velocity models (Figure 8.9) obtained from FWT also giving an idea on the delineation of finer details and of the major interface structures, which are clearly absent in travel time tomography models. Still, these FWT models are not sufficient to explain all the finer details present in the true model. This may be due to non-linearity of the problem, acoustic approximations, sufficient band-limiting, discrete frequency selection,and many inherently inexplicable problems. These models can further be improved by pursuing FWT with a greater number of iterations and by including higher frequencies. However, in reality, one does not know the subsurface structure, but they derive the velocity model from the acquired seismic data that are band-limited and may not be sufficient to resolve all of the finer details in the order of log scale, even one that performs the sophisticated FWT. Hence, we shall always get a relatively larger wavelength velocity model from field seismic data than the model obtained from the well log data.

Thus, taking all of these situations, 1D velocity models comparisons (Figure 8.10) and 2D velocity models (Figure 8.9), we see that the models obtained in case 2 and case 6 are in a good correlation with one another and maintain consistency in delineating the major structural features of the true model. We can draw some important conclusions in obtaining the optimum solution from the acoustic FWT. Irrespective of the initial model, the overlapping of the frequencies produces reliable results. As we discussed earlier, the inherent error in initial model, which was constructed from any other practical approach, will have an advantage in examining the capability of the algorithms rather than the smoothed version of the true model. Further, this explains the complexities in the application of acoustic FWT with real field seismic data during the delineation of the fine-scale details for different applications such as reservoir characterization, etc. This also suggests that the interpreters need to carefully look into the FWT models with the most prior information possible due to the well log details, which are confined to a single position in the profile.

8.5 Conclusions

Though the FWT has been applied to a limited number of field seismic data, the tool mostly lies in the theoretical domain to understand the intricacies during the multiscale imaging. We investigated the role of the starting model and the choice of frequency in acoustic FWT from elastic wide-angle synthetic data that have been generated for a realistic model, obtained from well logs. Since the FWT is highly computationally intensive, these kind of synthetic case studies are very useful in understanding the practical issues of successful applications to field seismic data. The study suggests that overlapping frequency groups with continuous intervals always produces the best possible results irrespective of the starting model. To demonstrate the capability of the FWT algorithm, many researchers use the smooth version of the true model as the starting model, and reaches to the solution without much difficulty. However, it may not be possible to have such a close starting model in all real-time scenarios. Further, it provides the flexibility to avoid the inherent propagation of errors in models, which are general in order to obtain by any practical approaches. In such situations, the traveltime tomography model or any other model that can be derived from the data itself can be used as the best possible starting model for demonstrating the efficacy of FWT algorithms. Otherwise, whoever is interested in using the smooth version of true model as the starting model, they should allow a sufficient smoothing filter by which they can build a model similar to one that can be obtained by any practical approach (e.g. first arrival/reflection traveltime tomography or stereo-tomography).

Although the FWT models show velocity-structures in more detail than TT models, some finer details are still missed in the FWT. This may be due to the application of acoustic modeling to realistic elastic data, the use of discrete set of frequencies, and most importantly, the inadequacy of wide-angle synthetic seismic data to resolve subsurface finer details in the order of log scale. One can recover the true model from the FWT of wide-angle synthetic seismic data only when the true model is close to seismic scale or greater than the log scale that has been used here. These types of studies play a key role in dealing with the highly computational imaging techniques where it takes a long time the best possible results within the timeframe. The usage of variable grid sizes is also useful in multiscale imaging, which has to be studied further to save computational time.

9

Travel Time Tomography of Seismic Data

9.1 Introduction

Six major tectonic elements are identified by Biswas and Singh (1988) from the western margin up to the Arabian abyssal plain. Mostly the western continental margin is comprised of five basins: Kutch, Saurashtra, Mumbai, Konkan, and Kerala. The Kerala-Konkan (KK) basin is the southernmost basin. The east-west trending Vengurla arch separates it from the Bombay offshore basin, and the Trivandrum arch defines its southern limit. The northern part of the basin is known as the Konkan basin, and the southern part is known as the Kerala basin. The Kerala basin's structural style is mainly controlled by an NNW-SSE trending fault parallel to the coastline and is mostly confined to the shelfal part and to a NNE-SSW fault oblique to the Miocene shelf edge and confined to the basinal part. The structural trend of the area is generally parallel to sub parallel to the coast (Chatterjee et al. 2006). Due to the existence of similarities in terms of sediments, morphology, and depositional characters, the Kerala and Konkan basins are considered as a single basin named the Kerala-Konkan basin. The western continental margin of India (Figure 9.1) evolved during the rifting of India from Madagascar and Seychelles during the late Cretaceous period. The adjoining of the Arabian Sea, the rifting and the seafloor spreading history of the western Indian margin, is well-documented in geophysical studies (Norton and Sclater 1979).

Since the Precambrian time, the following five major events have interposed the geological history of western India (Gombos et al. 1995):

- Subsequent rifting from Madagascar
- Reorientation of the regional drainage system from east-west to west-east and its effect on the sediment supply to the west coast

Active Seismic Tomography: Theory and Applications, First Edition. Kalachand Sain and Damodara Nara.
© 2023 John Wiley & Sons, Inc. Published 2023 by John Wiley & Sons, Inc.

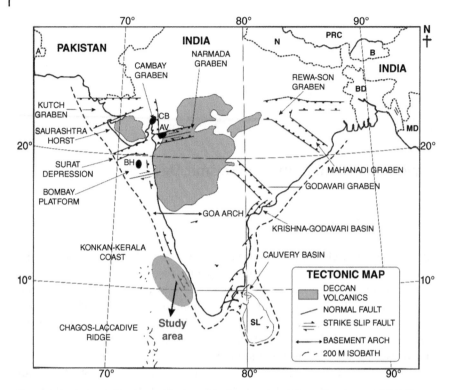

Figure 9.1 A generalized tectonic map of India reproduced from (Gombos et al. 1995). Locations of major hydrocarbon occurrences are indicated by ovals: AV = Ankleswar field; BH = Bombay High field; CB = Cambay field; A = Afghanistan; B = Bhutan; BD = Bangladesh; M = Myanmar (Burma); N = Nepal; PRC = People's Republic of China; SL = Sri Lanka.

- Deccan/Reunion mantle plume initiation with an associated uplift, extension, and subsidence
- Drift of India toward tropical latitudes and the development of a carbonate system
- Himalayan orogeny and resultant tectonic reactivation

According to the Directorate General of Hydrocarbons (DGH), the basin falls under prospective basin category-III and that thus far 15 exploratory wells were drilled in the basin and while none of them are producing, three of them are showing faint oil fluorescence. Over the past decades, numerous research groups are working in the KK offshore basin because the hydrocarbons in western continental margin of India are similar to Mumbai's offshore. ONGC acquired 2D seismic data in 2012 to understand the basin, and particularly to search for sub-basalt sediments (probably Mesozoic's) and basement configuration.

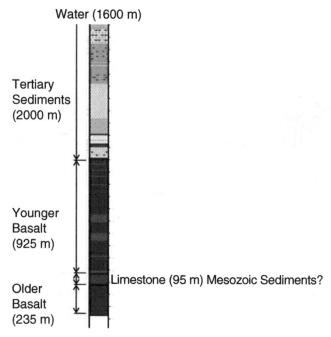

Figure 9.2 Litholog available nearby the SBN lines.

The litholog (Figure 9.2) at Well-A in the KK offshore basin shows a broad ~2.0 km thick Tertiary sediments, 925 m basalt (Paleocene Trap), 95 m Limestone (probably Mesozoic sediments) followed by another flow of basalt below 1.6 km of water column. The drilling of the well was stopped after 225 m into the second flow of basalt. The purpose of the wide-angle seismic experiment is to delineate the basement configuration and lateral extension of the Limestone formation. Here, we pursue the traveltime tomography of wide-angle seismic data to achieve the above goal.

9.2 Geological Setting

The Kerala and Konkan basins were divided with the Tellicherry Arch of basement high (Singh and Lal 1993). Searching for hydrocarbon reservoirs in complex volcanic and thrust-fold scenarios have always been a challenge to geophysicists. The KK offshore basin is one where we expect hydrocarbon prospects; however, the basin still does not show any prospect although the Bombay-High and Cauvery basin located in the north and south of the basin have been successfully exploited.

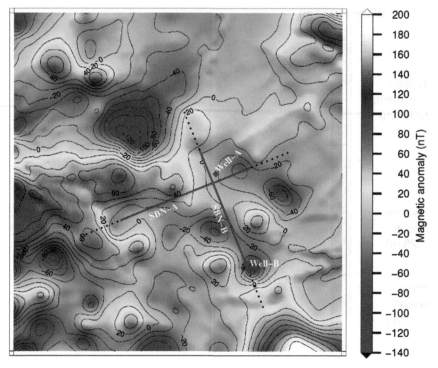

Figure 9.3 Total magnetic anomaly field map of the study region.

9.2.1 Magnetic Data

The magnetic anomaly map (Figure 9.3) of the study region is prepared with data available by Meyer et al. (2017) through the public domain https://www.ngdc.noaa.gov/geomag/emag2_download.html. The data are prepared with two arc-minutes resolution from the compilation of satellite, ship, and airborne magnetic measurements.

9.3 Data Acquisition

The ocean bottom seismometer (OBS) data were acquired in the western continental margin of India (Figure 9.1) by the Oil & Natural Gas Corporation (ONGC). The total acquisition along two SBN lines SBN-A (W-E) and SBN-B (N-S) covering ~400 LKM of wide-aperture seismic profiling using three component geophones and one hydrophone was carried out in the KK offshore basin in 2012. The data were recorded with a shot interval of 37.5 m and OBS spacing of 2.0 km along both the lines crossing each other. The shooting was carried out at least 40 km away from the end locations of

OBS positions. A total of 128 OBS stations were deployed over two SBN lines. The bathymetry varies from a minimum of 0.2 km to a maximum of 2.4 km along the lines of passing through two nearby wells (Figure 9.4). The detailed acquisition parameters are summarized in Table 9.1.

First arrival seismic refractions are the most certain and easiest to identify on a seismogram, and they provide more reliability on building the subsurface velocity structure. We use the first arrival traveltimes along each line individually in the KK offshore basin (Figure 9.4) to delineate the large-wavelength velocity structure from seismic tomography. Two representative OBS gathers and the picked first arrival traveltime data are superimposed on respective OBS gathers (Figure 9.5). The traveltimes were assigned with variable uncertainties of a minimum of 20 ms near the offset and a maximum of 120 ms at the far offset range. This uncertainty assignment is subjective

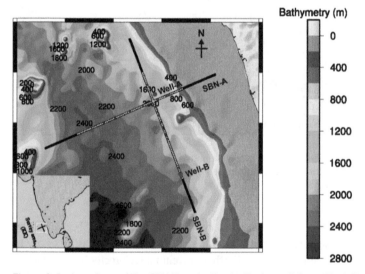

Figure 9.4 Locations of the SBN lines in Kerala-Konkan offshore. Black lines are MCS lines along which the SBN data have been recorded. Yellow dots are the locations of SBN, Red dots represent missing SBN. Blue solid circle is the location of Well-A, which is away from SBN lines, and the Red solid circle is the location of another well-B over and almost at end of the SBN-B line.

Table 9.1 Acquisition parameters along SBN lines in Kerala-Konkan offshore basin.

S.No.	Profile Name	Profile length(km)	Total No. of OBS's deployed	No. of OBS's retrieved	No. of OBS's lost
1	SBN-A	200	65	57	8
2	SBN-B	200	63	51	12

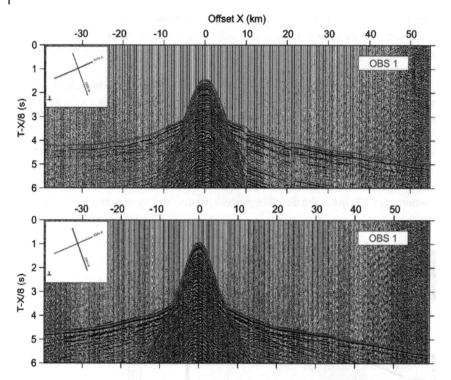

Figure 9.5 Representative OBS/SBN gathers along SBN-A (top) and SBN-B (bottom). The picked first arrival traveltimes are superimposed on respective OBS gathers. The gathers are plotted with a reduction velocity of 8.0 km/s.

and is based on ~1/3rd of the dominant cycle of the data as has been used by Sain et al. (2000). Assigning uncertainties to the picked arrival times is necessary to avoid over- and under-fitting the data. Further, traveltime reciprocity is also checked for all of the source-receiver pairs while the arrivals are picked for modeling (Zelt 1999).

9.4   Traveltime Tomography

9.4.1   Starting Velocity Models

Based on the proposed workflow presented in chapter 5, we derived the seven different OBS's bins used for 1D models (Figure 9.6) along the profile for SBN-A (three OBSs bins for the profile SBN-B) and then juxtaposed to construct the node based, minimum-structure, 2D heterogeneous models by correlating different 1D models along each SBN profile. The derived pseudo 2D initial models along each SBN line are shown in Figure 9.7. Each 1D velocity model is constructed by

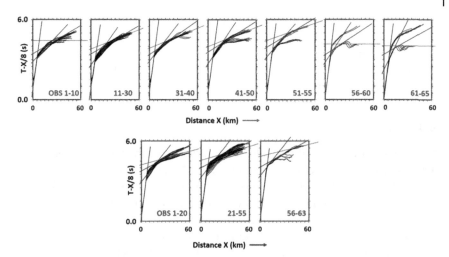

Figure 9.6 Different OBS's bins used for deriving 1D models with the intercept-time method along the profiles SBN-A (top) and SBN-B (bottom). The first arrival data are plotted with a reduction velocity of 8.0 km/s against the absolute distance.

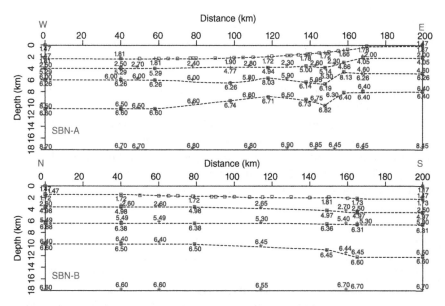

Figure 9.7 Starting models along SBN-A (top) and SBN-B (bottom) lines used for traveltime tomography. The defined velocity nodes are indicated by Blue dots, and Red color squares indicate the layer boundary nodes. The velocity values are noted at both the top and bottom.

finding slopes of straight lines for each segmented first arrival traveltime versus the absolute offsets compartments using the intercept-time method (Dobrin and Savit 1988). Since the intercept-time method gives 1D sharp velocity models, these are smoothed by applying simple moving average filter to allow seismic rays to pass at a maximum deep enough through the model.

9.4.2 Inversion

We parameterize the subsurface into 801×73 nodes for the forward problem and 400×36 cells in the inverse problem for each SBN profile. The approach (Zelt and Barton 1998; Zelt et al. 2003) of First Arrival Seismic Tomography (FAST) was employed to invert the first arrival traveltime data. The subjective free parameters included in the objective function [$\varphi(s)$] (Equation 9.3 in Chapter 6) are: the starting model, forward and inverse model grid spacing, initial trade-off parameter (λ_0) that maintains the overall data misfit, horizontal versus vertical smoothing factor (Sz), perturbation regularization weighting factor α, and number of iterations to solve the sparse linear system (Nolet 1987). In a strict sense, λ is not a free parameter, as it is systematically reduced with iterations from a starting value of λ_0 by the algorithm. The values of free parameters are varied to derive a minimum-structure simplest velocity model that is geologically most reasonable and that fits the data appropriately.

The initial trade-off is selected by trial and error, so that the RMS traveltime residual decreases by 10–30% after the first iteration by reducing the trade-off parameter (λ) with a factor (λ_r) of 1.41 ($\sqrt{2}$). This parameter helps control the inversion time in model space. The starting velocity models (Figure 9.7) are iteratively updated using the free parameters of the inversion process to derive the final model. At each iteration, traveltimes are calculated by forward modeling through the updated velocity model and new ray paths are simultaneously traced from all the source to receiver pairs. The bathymetry data among the lines were kept fixed during both the forward and the inversion stages with a constant water velocity of 1.47 km/s. The RMS traveltime misfit and the normalized χ^2 misfits are calculated by comparing the calculated traveltimes with the observed ones. These values corresponding to both initial and final preferred models are shown in Table 9.2 for

Table 9.2 Traveltime tomography parameters along SBN lines.

Name Of the profile	Dimensions				For initial model		For final model	
	Forward (grid size, km)	Inverse (cell size, km)	No. of iterations	Smoothing Parameter (s_z)	RMS Residual (ms)	Normalized χ^2 value	RMS Residual (ms)	Normalized χ^2 value
SBN-A	0.25 x 0.25	0.5 x 0.5	12	0.12	252.90189	14.2980	59.40583	1.0688
SBN-B	0.25 x 0.25	0.5 x 0.5	14	0.12	302.54633	19.7377	62.60102	1.0219

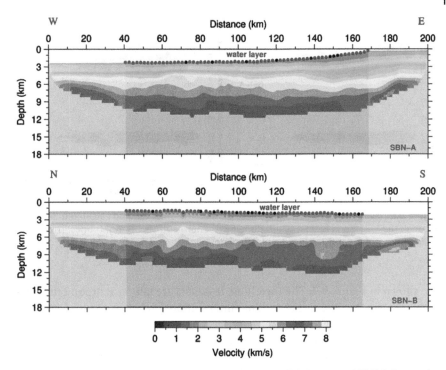

Figure 9.8 Final traveltime tomography models along SBN-A (top) and SBN-B (bottom). A Red color circle represents the OBS locations from which the data is retrieved and the Blue color circled represents the missed OBS locations. Gray color indicates the region which is not traced by the rays. Shaded region representing where the model is less controlled by the rays. White color indicates the water layer.

each profile, which we achieved after approximately 12 and 14 non-linear iterations for SBN-A and SBN-B profiles, respectively. We have used the smoothing factor (Sz) of value 0.12 for both lines, which is related to minimizing vertical versus horizontal model roughness and is selected subjectively as suggested by Zelt and Barton (1998). The final preferred velocity models obtained after tomographic inversion are shown in Figure 9.8. The modeling and inversion statistical parameters used for traveltime tomography along each SBN line are shown in Table 9.2.

9.4.3 Model Assessment

The derived models are reliable for interpretation only made after a necessary assessment that makes the interpreter confident about the models. Here, the models obtained from the tomography are assessed through different ways that includes ray path statistics, comparison of the traveltime data and their residual, etc. The traveltime residuals between observed and calculated data

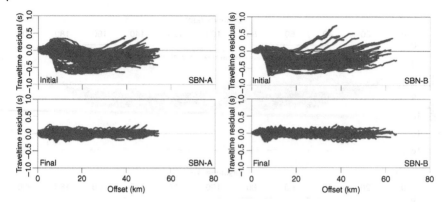

Figure 9.9 Data residuals (the difference between the observed and calculated data) through the Initial and Final models along SBN-A (left) and SBN-B (right) lines. The data residual is tremendously reduced and fallen around the zero level.

are shown in Figure 9.9. The traveltime residuals are falling at around the zero level compared to the initial residual information. The error reduced tremendously from very large range to a very acceptable range for the final preferred traveltime tomography models for both profiles the SBN-A and SBN-B. This specifies that the models are not biased due to the behavior of the symmetric nature about the zero position for the final traveltime residuals (Figure 9.9). Figure 9.10 also shows solid agreement among the calculated data through the final models (Figure 9.8) and the observed data for each profile. The rays traced through the final models are shown in Figure 9.11, which gives information on how well the model resolved through a particular cell/position at that depth location. Figure 9.12 represents the number of rays (ray hit count) passing through the cell at a particular location associated with final models (Figure 9.8). From all of these (Figures 9.9, 9.10, 9.11, and 9.12), it is clear that the derived final traveltime tomography models along each profile are well constrained and resolved during the inversion.

Figure 9.13 gives the reduction of the Chi-square value or the convergence history over non-linear iterations that occur in each profile with a corresponding trade-off parameter or lambda (λ, which is) used corresponding for the final tomographic models achieved. The λ is reduced with a reducing factor of 1.41 at successive iterations to stabilize the inversion without getting trapped into local minima and satisfying the linearization assumptions. We also have an indirect assessment for the models by comparing the velocity-depth function (Figure 9.14) at the cross-points of SBN lines, which show that the large-wavelength velocity models derived from traveltime tomography are reasonable. The velocity-depth function at the

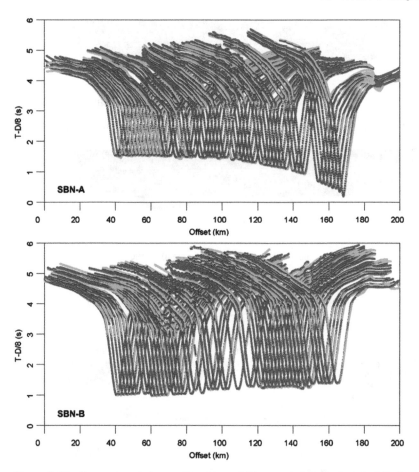

Figure 9.10 Comparison between the observed (Green color) and calculated (Red color) traveltimes for each OBS along the SBN-A (top) and SBN-B (bottom) lines. Traveltimes are plotted with a reduction velocity of 8.0 km/s.

location of Well-B along the SBN-B line also matches quite accurately with the high-frequency sonic velocity-depth function (Figure 9.14) derived from the log data. Hence, the models obtained from TT are well constrained and can be used for subsequent analysis and interpretation.

9.4.4 Results and Discussion

The traveltime tomography reveals the velocity variation from 1.47 km/s to 7.1 km/s along the SBN-A (W-E) and SBN-B (N-S) lines. These long-wavelength

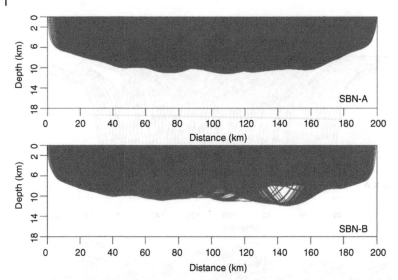

Figure 9.11 Raypaths through the final traveltime tomographic models along the SBN-A (top) and SBN-B (bottom) lines. The rays are displayed at the 5th ray only.

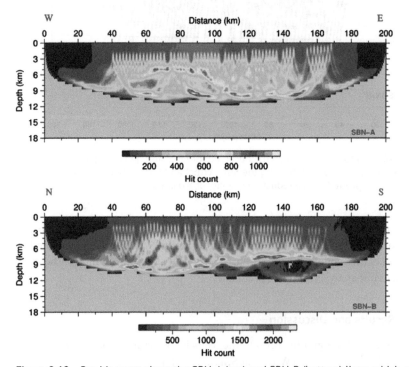

Figure 9.12 Ray hit count along the SBN-A (top) and SBN-B (bottom) lines, which gives the number of rays traced through each cell at a particular location of the model.

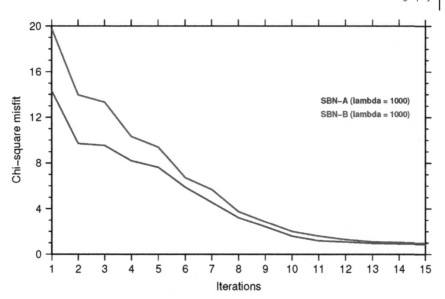

Figure 9.13 The reduction of the Chi-square value or convergence history over the non-linear iterations along each profile with the corresponding trade-off parameter or lambda (λ) used. After each iteration, the λ is reduced by a factor 1.41 to stabilize the inversion without getting trapped into local minima and satisfying the linearization assumptions.

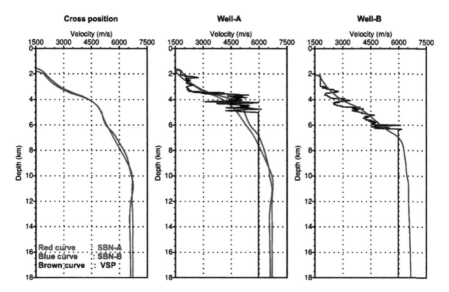

Figure 9.14 Left: Extracted velocity-depth functions at a cross-point of SBN-A and SBN-B lines. Middle: Comparison of extracted velocity-depth functions at projected well locations along both SBN lines with VSP velocity at Well-A. Right: Comparison of extracted velocity-depth function with sonic velocity at Well-B.

velocity models are broadly characterized into five layers including the water column. The major geological formations of each layer are Eocene, Paleocene Trap, and Cochin formation or Limestone, Older trap, and basement, respectively. Due to the limitation of the first arrival traveltime tomography and the presence of thin Limestone beneath the Paleocene Trap or basalt, as evidenced from the VSP logs, the models derived from TT are not enough to delineate the thin layer and its lateral along the lines. The basement with the velocity range of 6.3–6.5 km/s shows the maximum depth ~10.0 km.

9.5.5 Conclusions

The long-wavelength velocity models are derived from the first arrival traveltime tomography of the wide-angle OBS data. The model shows the basement configuration along each SBN line. Though the traveltime tomography does not show any low-velocity signature corresponding to the sub-basalt Mesozoic sediments, the derived models are used as starting models for the state-of-the-art full waveform tomography to derive the finer details of the subsurface, including the thin sub-basalt sediments. These models are also very useful in avoiding the cycle-skipping problems of full waveform tomography. If the age of older trap is established by the geochronological studies, it will open a new direction for understanding the evolution of Mesozoic basin in the KK basin.

10

Full Waveform Tomography of Seismic Data

10.1 Introduction

The Kerala-Konkan (KK) basin in the southeastern part of the western continental margin of India has gained much interest because of its anticipated hydrocarbon potential. The basin lies to the south of Bombay's offshore basin, India's major hydrocarbon producer, and is adjacent to the east coast Cauvery basin where hydrocarbons (DGH) are established. The estimated resource base of the basin is in the order of 660 million tons, according to ONGC estimates (https://mopng.gov.in/en/exp-and-prod/conventional-hydrocarbon). The KK basin covers an area of about 1x69,000 sq. km and many exploration wells were drilled to identify the existence of carbonates/clastics of the tertiary in early 1980s, but none of them succeeded. The basin's large extent and its closeness to the prolific, producing Mumbai offshore basin in the north have made it significant for hydrocarbon exploration. But the exploration so far has not yielded any discovery, raising some doubt as to the existence of a petroleum system in the basin. The KK basin is one of the vast, unexplored deep water frontier regions. Therefore, the optimistic researchers take another direction in searching rift-systems, igneous-lows, and tectonic activities, which are reasonable for hydrocarbon occurrences and to understand the basin's evolution.

This chapter presents results from the full waveform tomography of wide-angle ocean bottom seismic (OBS) data and emphasizes delineating sub-basalt sediments, which couldn't be made by traveltime tomography.

10.1.1 Gravity Data

The Bouguer gravity anomaly map (Figure 10.1) was prepared using data (Barthelmes and Köhler 2016) available in the public domain http://icgem.gfz-potsdam.de.

Active Seismic Tomography: Theory and Applications, First Edition. Kalachand Sain and Damodara Nara.

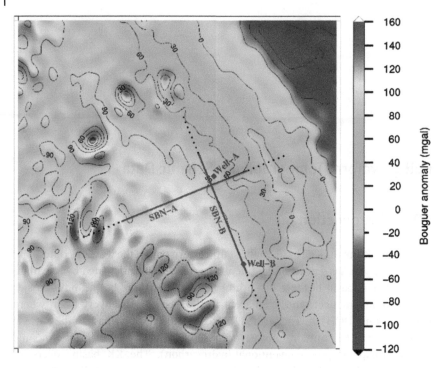

Figure 10.1 Bouguer gravity anomaly map in the study area showing the seismic SBN lines and locations of Well-A, Well-B.

The map depicts the anomaly as increasing toward the Southwest (SW) from the Northeast (NE) following the general trend of bathymetry. The anomaly falls in the positive range due to the increase of attraction that is primarily filled with dense basalts (Deccan basaltic flow of age 65 Ma).

10.2 Marine Seismic Survey

ONGC acquired the wide-angle OBS data in the KK offshore especially for establishing the Mesozoic basin. Here we perform FWT along two lines, SBN-A and SBN-B (Figure 10.1), with the aim of delineating the basement configuration and the sub-basalt sediments for hydrocarbon explorations. The data acquisition parameters are given in Table 10.1. Representative OBS gathers along SBN-A and SBN-B lines as shown in Figures 10.2 and 10.3, respectively, along with the first arrival traveltime data. The OBS's are deployed with an interval of 2.0 km and a shot interval of 37.5 m. The data were recorded with an 8 ms and a 4 ms sampling interval along the SBN-A and SBN-B, respectively. The starting models for FWT are derived from the first arrival traveltime tomography.

Table 10.1 Data acquisition and the subsurface model parameterization in Kerala-Konkan offshore basin.

Parameter	SBN-A	SBN-B
Model dimensions *(km)*	128 × 18	125 × 18
Total No. of SBN's deployed	65	63
No. of SBN's retrieved	57	51
No. of SBN's lost	8	12
Grid size *(m)*	12.5	12.5
Model dimensions in no. of grids *(nx, nz)*	(10241, 1441)	(10001, 1441)
V_{min}, V_{max} *(m/s)*	1470, 7000	1470, 7200
Sampling interval *(ms)*	8	4

10.3 Full Waveform Tomography

As the crustal velocity structure is exploited using the traveltime tomography of selected phases from the whole seismograms, the small-scale structures or features cannot be resolved. It is to be stated that the resolution of the traveltime tomography is in the order of the Fresnel zone, whereas, the entirety of subsurface information (seismic velocity, density, attenuation, and anisotropy) is contained into whole seismograms and can be exploited by full waveform tomography (FWT) in which the resolution is in sub-wavelength order is much smaller than the Fresnel zone.

We applied the Visco-acoustic FWT in the frequency-domain using the classic gradient method (Pratt et al. 1998) in which the full-wave equation is used rather than a one-way wave equation that ignores backscattering and wide-angle effects. The gradient of the objective function is calculated using the adjoint method by back propagating the residuals and correlating with the forward wavefield to avoid explicit computation of the partial derivatives. Theoretical details on FWT are already given in the first section.

10.3.1 Initial Models

Since the FWT uses gradient methods, we always start with a very good initial model, and the long-wavelength velocity model, as derived from the traveltime tomographic approach of Zelt and Barton (1998), serves the purpose. As the inversion progresses from low-frequency to high-frequency, the minimum wavelength details are incorporated in the solution in multiscale imaging fashion. The final traveltime tomography results along both SBN-A and SBN-B lines and are

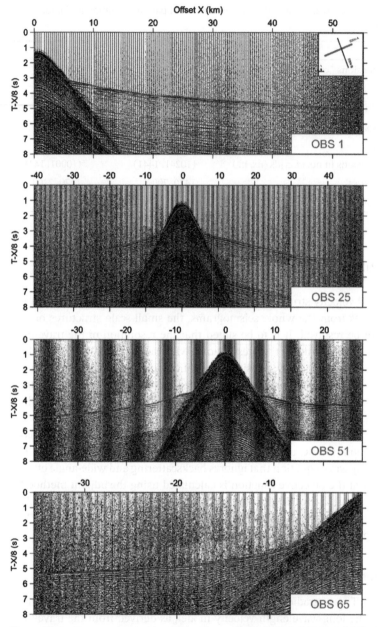

Figure 10.2 Four representative OBS gathers: OBS1, OBS25, OBS51, OBS65 along SBN-A profile, plotted with a reduction velocity of 8.0 km/s and shown with handpicked first arrival times.

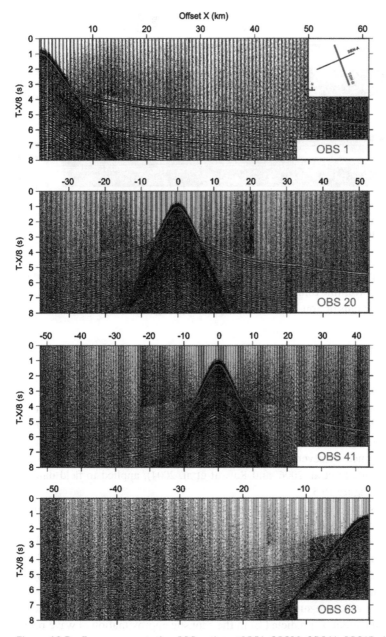

Figure 10.3 Four representative OBS gathers: OBS1, OBS20, OBS41, OBS63 along SBN-B profile, plotted with a reduction velocity of 8.0 km/s and shown with handpicked first arrival times.

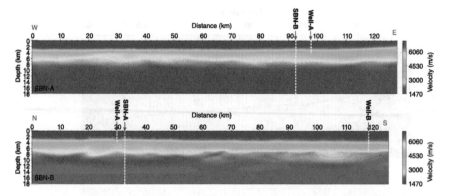

Figure 10.4 Velocity models derived from the traveltime tomography of the first arrival seismic data along SBN-A (top) and SBN-B (bottom) lines respectively. The downward arrows show the cross and well positions.

shown in Figure 10.4. The details are presented in the previous chapter. The long-wavelength traveltime tomography models are also helpful in mitigating the quite common problems of non-linearity and cycle skipping in FWT.

10.3.2 Data Pre-processing

Since we applied the frequency-domain FWT in acoustic approximation, data pre-processing is necessary to improve the S/N ratio and to mitigate or exclude the deep reflection arrivals and propagation effects in data that cannot be predicted from acoustic approximation. For the successful application of FWT, numerous studies are designed for the pre-processing steps. A sequence of pre-processing flow, as designed by several workers (Brenders and Pratt 2007a, 2007b, Kamei et al. 2013, Operto et al. 2006, and Ravaut et al. 2004), applied to field data, is described in Figure 10.5.

A bandpass filter was applied with frequencies 1–2–20–25 Hz to exclude high-frequency components so that error propagation will be mitigated during the inversion from low to high frequencies (Figure 10.5). Then noisy traces are killed. Finally, a variable time window was applied depending on the water depth with a minimum 0.5 s at shallow water and a maximum 2.5 s at deep water (Figure 10.6). The time window was applied to exclude non-acoustic arrivals such as the P-SV converted arrivals, free-surface multiples, and early ambient noise contained in the data that cannot be predicted from the acoustic approximation of wave propagation. Representative pre-processed data for selected OBS are displayed in Figure 10.6. The data were subsequently Fourier transformed, and selected frequency components were extracted to form a mono-frequency dataset for FWT.

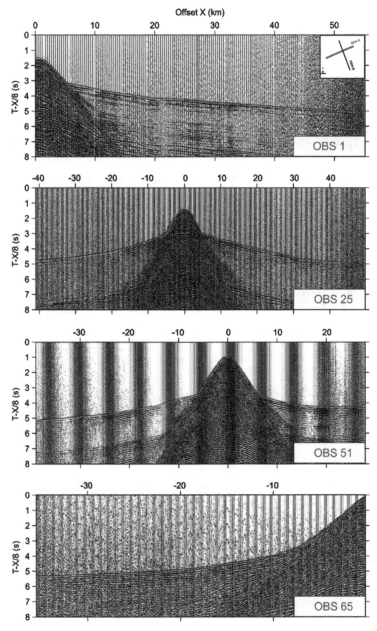

Figure 10.5 Examples of bandpass filtered OBSs data along SBN-A and SBN-B (next page) lines.

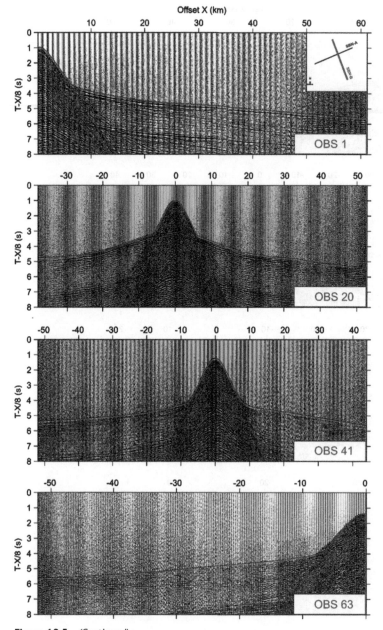

Figure 10.5 *(Continued)*

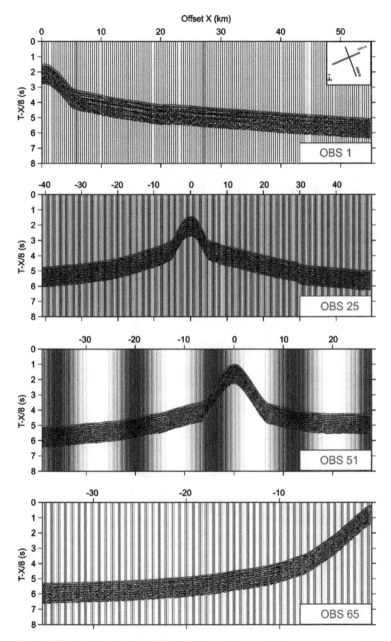

Figure 10.6 Pre-processed OBS gathers after muting applied to the gathers shown in Figure 6.5 along the SBN-A and SBN-B lines. The pre-processed data are transformed into the frequency-domain for full waveform tomography.

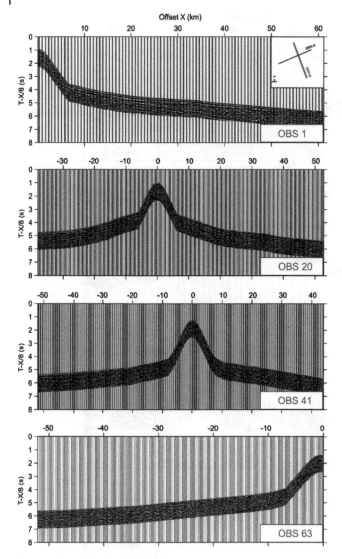

Figure 10.6 *(Continued)*

To stabilize the inversion and to fall the solution in the basin of attraction, which is close enough to the global optimum, inversion is to be carried out starting from low-frequency and by subsequently switching over to higher frequencies in the multiscale approach. There are two choices of the discrete frequency selection for frequency-domain FWT. In the first choice, frequencies are sampled at a uniform interval, whereas, in the second choice, the frequencies are optimally selected at discrete levels by the criteria designed by Sirgue and Pratt (2004). This criterion will produce a discrete set of frequencies so that it has full, continuous, vertical wavenumber coverage. We followed the first choice, which is extensively applied to crustal scale FWT applications (Operto et al. 2006; Ravaut et al. 2004). Further, it guarantees the correct sampling of the wavenumber spectrum by the uniform frequency sampling (Malinowski and Operto 2008).

10.3.3 Application of Full Waveform Tomography

We have chosen the starting frequency of 2.0 Hz to compensate for the long-wavelength features lying within the traveltime tomography. This possible low-frequency is also useful to overcome the cycle skipping problems that occur during FWT. The tomography was performed at every 0.5 Hz frequency interval up to a maximum frequency of 20.0 Hz. This frequency interval is sufficient for the fine sampling range of vertical wavenumbers in the subsurface image as discussed in section 8.3.3.1. A total of 37 frequencies were formed into 36 groups with two frequencies at each group and one overlapping frequency between the groups. The representative spectrum of the data is shown in Figure 10.7. All the statistical information is shown in Tables 10.1 and 10.2. The traveltime tomography models shown in Figure 10.4 (Sain et al. 2017) are used as starting models for FWT. The

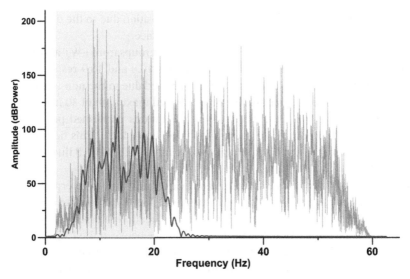

Figure 10.7 Representative spectrums of the data at selective traces. The Pink curve for the raw data and Blue curve for the pre-processed data. The shaded region describes the range of frequencies that has been used for full waveform tomography.

Table 10.2 Full waveform tomography parameters in Kerala-Konkan offshore basin.

Parameter	Description
Frequency range	2–20 Hz
No. of frequency groups	36
Max. no. of iterations/group	30
No. of frequencies in each group	2
Frequencies within group	Group 1: 2, 2.5 Hz;
	Group 2: 2.5, 3 Hz;
	Group 3: 3, 3.5 Hz;
	....
	Group 36: 19.5, 20 Hz

FWT was performed in a multi-scale approach starting from low to high frequencies by taking the final model of the previous frequency group as the starting model for the next frequency group. We have allowed 30 non-linear iterations at each frequency group to update the model along with the stopping criteria, where the model update is negligible as deterministic fashion. We kept the constant water velocity of 1.47 km/s throughout the inversion by incorporating the bathymetry interface into the background model. We also excluded the data at the near offset range of 10.0 km around each OBS position due to the dominant contribution of large amplitudes at the sea-floor interface.

The velocity models derived at different frequency groups from FWT along the SBN lines SBN-A and SBN-B are shown in Figures 10.8 and 10.9 respectively. The bandpass filtered spike source signature was used due to the non-availability of the original source signature. We have allowed maximum of 30 iterations for each frequency group. But, the maximum iterations performed per group with respect to the stopping criteria are shown in Figure 10.10. This figure also gives that the number of non-linear iterations are more in the SBN-B line than in the SBN-A line.

The absolute velocity models of the subsurface are difficult to interpret with the broad velocity range 1.47–7.2 km/s. Further one has to explain the absolute velocity models cautiously due to an inherent propagation of errors caused by acoustic approximations. Thus, we take the difference between the FWT results and the initial model, which is called the perturbation model, to interpret contribution significance by FWT in delineating subsurface reflectors. The perturbation models are shown in Figures 10.11 and 10.12 for the SBN-A and SBN-B lines respectively.

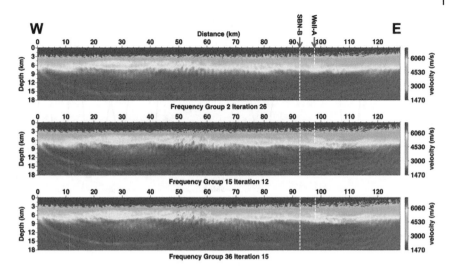

Figure 10.8   Velocity models derived from full waveform tomography along the SBN-A line at different frequency groups. The downward arrows show the cross and well positions.

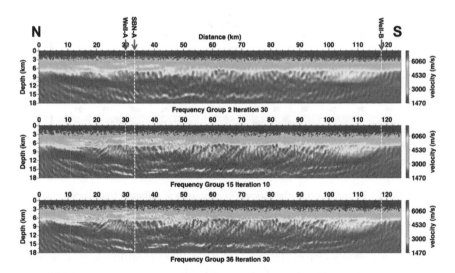

Figure 10.9   Velocity models derived from full waveform tomography along the SBN-B line at different frequency groups. The downward arrows show the cross and well positions.

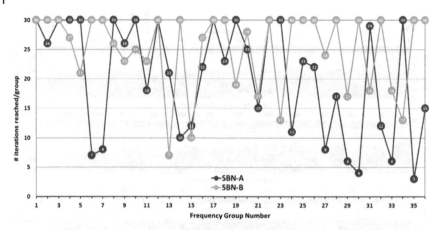

Figure 10.10   Number of iterations at different frequency groups along both SBN lines.

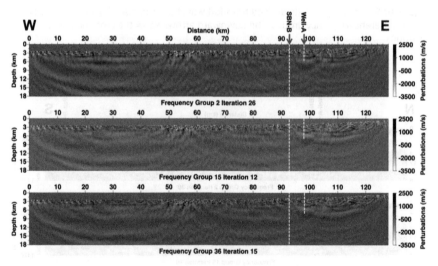

Figure 10.11   Perturbation models shown in gray scale at different frequency groups along the SBN-A line.

We made a comparison of 1D profiles at three different positions: cross position between the profiles, and the Well-A and Well-B locations, which are displayed in Figure 10.13. Since the Well-A is away from the SBN lines, we have extracted 1D models at projected positions based on the 2D FWT results along both lines to correlate with the geological formations observed in the well. Figure 10.14 shows 1D velocity depth profiles extracted from the TT and FWT velocity models at every 10 km interval along the SBN-A and SBN-B lines respectively. This explains the

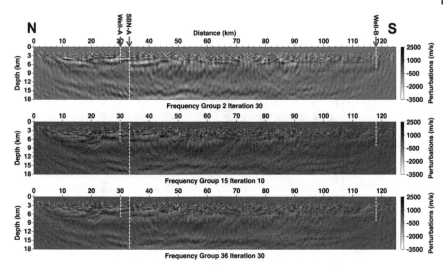

Figure 10.12 Perturbation models shown in gray scale at different frequency groups along the SBN-B line.

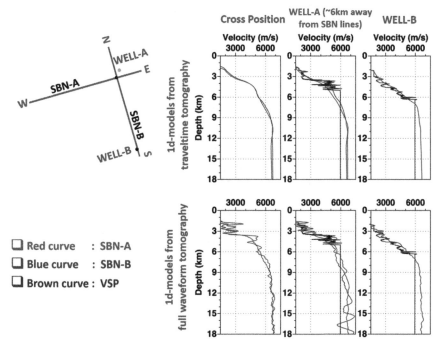

Figure 10.13 Comparison of 1D velocity depth profiles at three different positions: Cross, Well-A and Well-B locations.

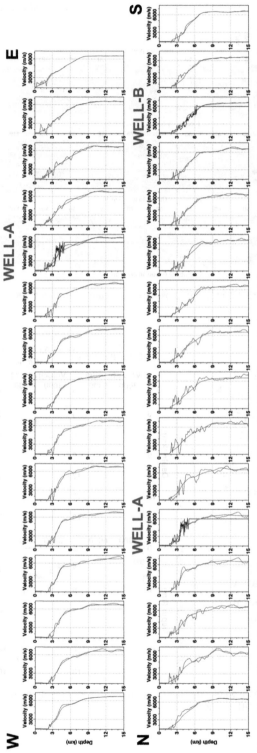

Figure 10.14 1D velocity depth profiles extracted from the TT (Red) & FWT (Blue) velocity models at every 10 km interval along the SBN-A (top) and SBN-B (bottom) lines. The logs are shown in the Black color.

variations of the models along the profiles; the finer details are observed in the shallow parts rather than in deeper portions. Further, to demonstrate the fine-scale velocity structure derived from the FWT, we zoomed in on the results along the SBN-A lines that were between 90 to 100 km and the results along the SBN-B lines thtat were 25 to 35 km, respectively (Figure 10.15). The low-velocity zone (LVZ) indicates the limestone formation, probably the Mesozoic sediments.

10.3.4 Computational Resources

We performed the FWT on the Linux platform server with four nodes having a total of 64 cores that were populated as 16 core Intel Xenon 2.4 GHz processors per node. It takes nearly 45 days to complete FWT for all 36 frequency groups along each SBN line. The model is parameterized with 10241×1441 and 10001×1441 finite-difference grids of step 12.5 m along SBN-A and SBN-B respectively. The gradient of the objective function is calculated in the decimated model with 20 and 10 grids (i.e., 250 m and 125 m) in depth (z) and distance (x) direction along each profile. Once the gradient is computed on the coarse grid model, it can be interpolated back to the fine grid at the original sampling rate 12.5 m.

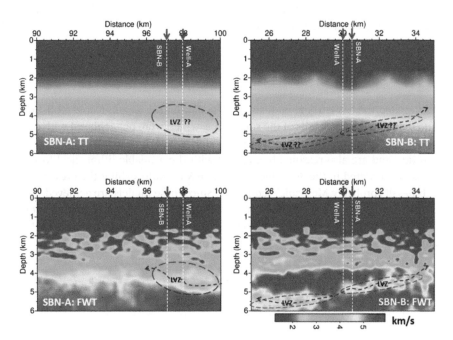

Figure 10.15 Zoomed portion of the FWT models (bottom panel) showing LVZ (Mesozoic sediments) below the basalt (Paleocene Trap) along the SBN-A (left) and SBN-B (right) lines. The finer details including the LVZ are not observed in the traveltime tomography models (top panel).

10.4 Discussion

From the geological history of the study region, the maximum potential clastic type of sedimentary rocks is expected rather than carbonate rocks. Further the clastic rocks exhibit the P- wave velocity range of 2–4.9 km/s. This has been reported from the East-Siberian Sea as a maximum of 4.7 km/s at an even greater depth of 12 km. Again, the basaltic rocks are also characterized with the higher velocity range of 4.7–5.8 km/s (Eccles et al. 2009, 2007; Sheriff and Geldart 1995) and are also conformed from the wells (KK-1: 4.55 km/s; KK-DW-17–1: 5.1 km/s; KKD-1A: 5.68 km/s) drilled through the Deccan formation in the surrounding region. Therefore, The Mesozoic sedimentary deposits, if present below the basalt flow, should show the low-velocity. The LVZ along the SBN lines as observed in the zoomed version of the final models (Figure 6.17) may thus indicate the sub-volcanic or sub-basalt Mesozoic sediments. Hence, the FWT is more capable of delineating low-velocity layers, which couldn't be delineated with conventional traveltime tomography. Further, the sub-basalt lithology constraint in forming the low-velocity zone is unpredictable with the crystalline basement observed in continents, and it also represents the Mesozoic sequence in the form of intruded-sills from pre-breakup of the basin. This is one of the very important conclusions we have drawn from the FWT results.

10.5 Conclusions

We applied the FWT to the wide-angle OBS data in the Kerala-Konkan offshore and delineated finer details of the subsurface, which were not achieved with conventional traveltime tomography. The models are well correlated at the cross position and are also reasonably matched with the available lithological formations encountered in the available well. The FWT results exhibit the presence of sub-basalt (probably Mesozoic) sediments. Since these models are achieved with a bandpass filtered spike source rather than the actual source due to its non-availability, we are sure that the models can be further improved by using the actual source signature.

11

Advanced Seismic Processing Using Tomographic Results

11.1   Introduction

Classical data acquired for oil and gas exploration has a maximum source receiver distance of order of 4 km that was made for near vertical incidence angles (Yilmaz 2001). If we want information beyond the critical angle of incidence, we need wide angle data, which means a maximum offset of the order of some tens of KM (Sheriff 2002). Using the wide angle seismic data, we can derive more information than we can from the primitive near vertical incidence data (Colombo 2005). It is observed that the strength of the seismic amplitude near the critical and post critical angle of incidence are higher than the amplitude near the vertical range (Aki and Richards 1980; Sheriff and Geldart 1995). Also, the large offset reflection can give better velocity information than a shorter one. Deep targets imaging is not possible using old near vertical incidence range techniques. Now, the oil and gas industries have required streamer length 12 km for acquiring wide angle seismic for deeper targets (Colombo 2005). A representative CDP gathers long offset seismic data from Kerala Konkan basin is shown in Figure 11.1. There are two types of analysis are being done for wide angle data: one that involves seismic data processing, and the other, inversion. Seismic data processing involves several steps for obtaining a promising image of subsurface (Robein 2003). Existing seismic imaging techniques are developed for the near vertical range of the incidence angle (Sheriff and Geldart 1995; Yilmaz 2001), which has certain assumptions; thus, it may not be appropriate for long offset data processing. Valuable parts of wide angle data can be discorded if we adopt a conventional processing flow (Yilmaz 2001). Wide angle data have deep target information; to get that information, we must include the wide angle amplitude in our stacked data. We should look past conventional techniques to alternative schemes for deconvolution, multiple removal, velocity analysis, muting, and move out correction.

Active Seismic Tomography: Theory and Applications, First Edition. Kalachand Sain and Damodara Nara.
© 2023 John Wiley & Sons, Inc. Published 2023 by John Wiley & Sons, Inc.

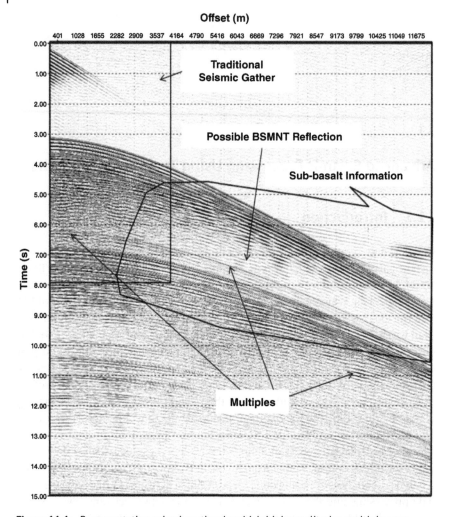

Figure 11.1 Representative seismic gather in which high amplitudes multiples are masking the reflection phases.

Problems that we encounter while processing long offset data are absorption of high frequency components, multiple reflected energy, converted wave, and stacking velocity analysis. Reflection amplitude from deep targets are weak in strength, which can be masked by multiples. The most desirable wide angle amplitudes coming from deeper targets may get masked by the tails of reflection hyperbola of shallow reflectors (Hanssen 2005). Near offset signals are strongly contaminated with multiples while far offset signals are contaminated with background noises. Since the far offset signals are not contaminated with long period multiple, multiple removal is relatively easy in wide angle range (Fig. 11. 1).

As the quality of the stack is very sensitive to velocity, velocity analysis with a small offset is very difficult. "Though there are good velocity information in the long offset gathers, ambiguities and complexities associated with the near-critical angle of incidence makes the velocity analysis for these wide angle data most challenging stage in processing" (Hanssen 2005). Imaging sub-basalt sediment is one of the concerns of the oil and gas industries. Problems with the sub-basalt exploration are complex due to the structure of the layers; successive lava flow can introduce heterogeneity within the basalt layer. This heterogeneity may cause scattering and diffraction of seismic energy from basalt top. The main concern regarding data processing in a region of volcanic flow is imaging reflection from base of basalt and, most importantly, basement reflection (Jones 2003). The interface between the base of basalt and top of sediments can be imaged easily as there is a very high contrast in impedance. The huge impedance contrast at this interface results in backscattering. A major part of incident seismic energy is reflected back at this interface (Ziolkowski et al. 2003), and a very small part of energy is transmitted. Transmitted energy is not that enough to go at a deeper layer come back to the receivers. Due to the huge reflection at base of the volcanic layer, reflections from this interface are so strong that they can give rise to strong basalt base multiples. Strong basalt base multiples can mask the weak short offset of primary reflections. These are some of the problems associated with sub-basalt imaging that make it challenging even after so much advancement in techniques (Ziolkowski et al. 2003). Some of these imaging issues can be handled by using wide angle seismic data. About 54% of oil and 44% of natural gas in the global hydrocarbon reserves are expected to be found in Mesozoic sediments. India has a huge 390,000 sq. km area of Mesozoic basins, which are considered frontier basins by the Indian Oil and Natural Gas Corporation. These basins are mostly covered by a basaltic layer (Deccan Trap) of late Cretaceous age. Indian Oil companies working on these frontier basins are very interested in the structures below the basaltic trap. CSIR-NGRI has already demonstrated the presence of low velocity Mesozoic sediments below the Deccan flow in the Saurashtra, Kutch, and Deccan Syncline regions with wide-angle seismic experiments (Sain et al. 2002a, 2002b, Sain and Kaila 1996, Sridhar et al. 2009, Murty et al. 2016, and Prasad et al. 2013). These velocities have no insight into structural details essential for oil and gas exploration in the locations of the Kerala Konkan (KK) basin south of the India's largest producing oil field of Mumbai's offshore basin, and near the east coast carvery basin. As discussed, velocity analysis is one of the most challenging tasks in processing long offset reflection seismic data, but a fair velocity structure, this can be found alternatively, i.e. with traveltime tomography of SBN data that is acquired along the long offset seismic line. This velocity can reduce our efforts considerably in imaging Mesozoic and basement reflections from long offset data.

11.2 Filter

The sampling interval of conventional seismic recording is 2ms, which is the nyquist frequency 250 Hz. The maximum frequency of 250 Hz is recorded just to avoid the aliasing because the 100 Hz frequency is enough for seismic experiments. That is why despite recording 2ms data, we resample it to 4ms and apply a 120 Hz high cut filter, as we are sure that there is no aliased noise in raw data. We have used one of our own filter designs that utilises the source signature extracted from seismic data in such a way that it can do minimum phase conversion, high cut filtering, and ghost notch filtering (Maslen 2013). The result of filtering on a shot gather is shown in Figure 11.1.

11.3 Denoise

There are three possible ways to remove linear noise: front muting, FK filtering, and Tau-P filtering. Front muting information from the far offset cannot be used, which is very useful for velocity picking and imaging. They are not visible because they get masked under strong linear noise. If we want to utilise them for offset data, we should look for alternative methods to remove these noises. Noise suppression in the Tau-P domain or the FK domain can be used; we have chosen FK domain as it can be combined with removing swell noise in the FK domain. To avoid spatially aliased components in the FK domain, we have used interpolated data at a small receiver interval for the FK transformation. After Denoise data is resorted at the original receiver interval, the result of Denoise in shot domain can be seen in Figure 11.2. We have also applied the FK filter in the CDP domain to remove the residual noise result, which can be seen in Figure 11.3.

11.4 Demultiple

Data processing in the area covered by the volcanic trap is complicated because of its heterogeneous nature, high reflectivity, and ability to generate multiples. Multiples of these horizons and seafloor multiples are much stronger than the original, deeper reflectors. Due to the heterogeneous nature and high reflectivity frequency, the containers and energy of deeper reflectors are already very poor. High energy of multiples makes imaging deeper layers more difficult; single multiple elimination technique may not work well. For this we need a comprehensive strategy for multiple removals. In the present study, we have shown a combination of processing sequences to attenuate the multiples in the KK basin complicated by

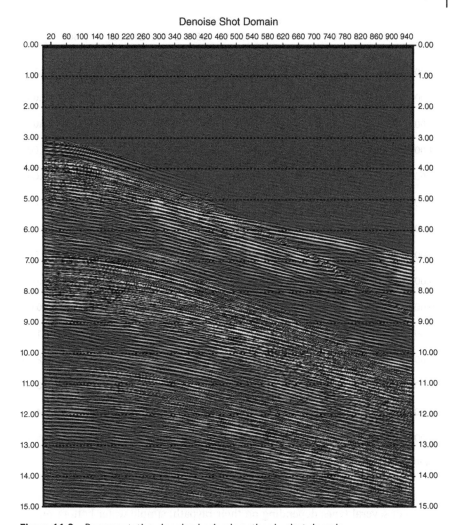

Figure 11.2 Representative denoised seismic gather in shot domain.

the basalt trap. Figure 11.4 shows the common offset section before eliminating multiple energy and shot gathers from two shot locations shown by arrows. Multiples have much stronger energy than primary, a huge contrast in impedance at the seafloor, and trap top enables these to generate multiples. Shallow trap top may complicate the multiple reduction more than the deeper trap top because the trap top multiple occurs a little early. In the left part of the section with seafloor multiples, basement reflectors and trap top multiples are very close to each other, while in the right part of the section, they are well separated. It can be explained

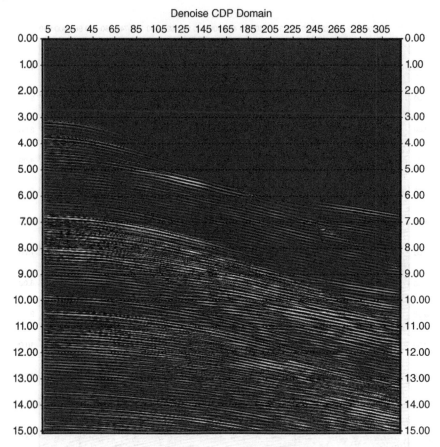

Figure 11.3 Representative denoisedseismic gather in CDP domain.

due to the shallow occurrence of the trap top in the left part of the line. Multiples can be of several types depending on their raypath, simple water bottom multiples, peg-legs, and inter bed multiples (Kurin et al. 2015). In marine seismic data, most of the multiples are simple water bottom multiples and are water layer reverberations; there are well known techniques to reduce these (Wiggins 1988). For the data, which is heavily contaminated by multiples, we should use other modern techniques for multiple elimination.

a) **SRMA:** Surface related multiple elimination technique (Berkhout and Verschuur 1997) is two step, multiple elimination procedure. In this process, we first model the multiples' energy using the parameter from the gather. In next step, this multiple model is adaptively subtracted from the real gather. Several iterations of SRME can be used if the water depth is very shallow.

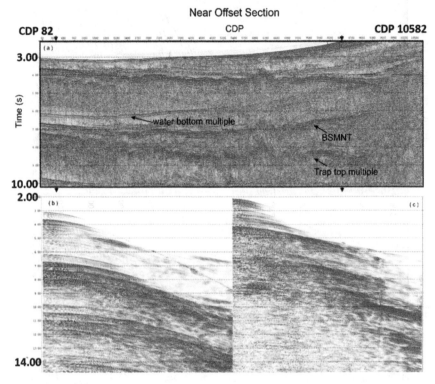

Figure 11.4 Top: Representative common offset section before eliminating multiple energy. Bottom: Shot gathers from two shot locationshown by arrows.

Primaries can be contaminated by second and third order multiples also. We have used a single iteration of SRME to the gather because only first order multiples can contaminate the primary reflection. Figure 11.5 shows SRME result for the gather from the left hand side of line; results show that SRME alone cannot suppress dipper multiples effectively, though, the shallow reflections have some improvement for effective multiple attenuation, which is a combination of multiple removal technique is required.

b) **RADON:** Velocity filtering may be the other option for multiples elimination; in general, increasing depth wave velocity in sediments increases while the multiples of shallow reflectors will have low velocity. It suggests that move out curves of multiples will have more curvature than high velocity deep reflectors and multiples can be separated from the primary with move out. Linear radon transform is useful for multiple attenuation in the large offset region. While parabolic radon transformation can effectively suppress the multiples in mid to large offset regions, to get a better result, we can use these in combination

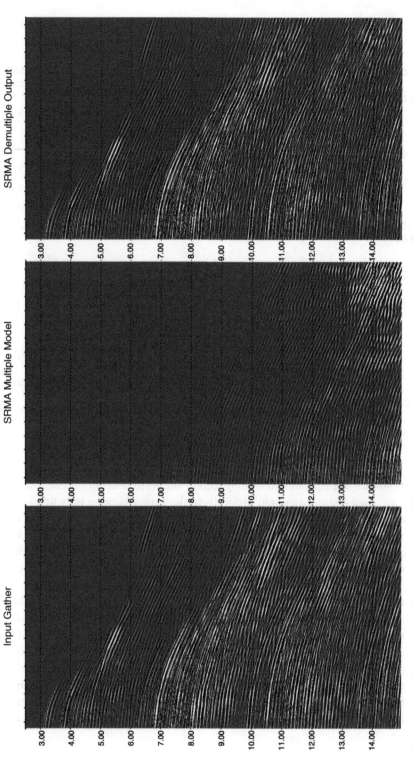

Figure 11.5 Illustration of SRMA. The output of SRMA enhancesthe signal strength by attenuating the energy of unwanted multiples.

(Kurin E. et al. 2015). For parabolic radon transform data, NMO corrected with the velocity should be a little lower than the actual velocity. Doing this primary reflection will get over corrected and multiples remains under corrected. When this NMO corrected data is used in parabolic radon transform, multiples will show positive slowness while primaries will show negative slowness. Energy concentrated in negative slowness regions in parabolic radon transform domains may be muted, and inverse parabolic radon transform is applied. Inverse NMO correction using the same velocity is applied to data to get multiple attenuated gather. Figure 11.6 shows the radon Demultiple result for the gather results shows a significant elimination in multiple energy. While multiple elimination in parabolic radon transform a wide pass band of slowness is applied in order to prevent the reflection from basalt layer and keep a scope for future velocity refinement. Using wide pass for slowness in parabolic radon transform domain there is always a scope for remnant multiple energy that can suppressed after pre-stack depth migration using the same radon Demultiple technique.

11.5   Post Migration Demultiple

The normal seismic section is prepared by correcting the source receiver offset to a zero offset (NMO) and stacking them. Doing this, we have not we have ignored the seismic raypath and diffractions arises due to discontinuities (faults and other heterogeneity). As a result, when we look at the conventional stack section, all of the dipping events will appear to be dispositioned from the actual position and the diffractions may contaminate the real reflection. Migration is the process that removes the wave propagation effect from the data, i.e., migration tries to bring receivers and sources at a common reflection point so that actual source generated signal can be recorded. So, migration can solve the issues of dispositioning events and diffraction due to sharp discontinuity and to intrusions. Migration repositioning of the reflectors at its actual lateral position and also the collapse the of the diffractions enhance the lateral resolution. Migration needs some velocity model as an input; if the velocity model is complex, it needs more computation. Pre-stack migration requires more computational time than post stack migration, but the advancement in computation speed has largely replaced post stack migration with pre-stack migration. Depth migration has the great ability to deal with the lateral velocity variation in compression to time migration. Kirchhoff's migration is the most popular migration technique, which is a kind of diffraction summation method that also incorporates obliquity, spherical spreading, and wavelet shaping factors (Schneider 1978). Time migration uses the RMS velocity

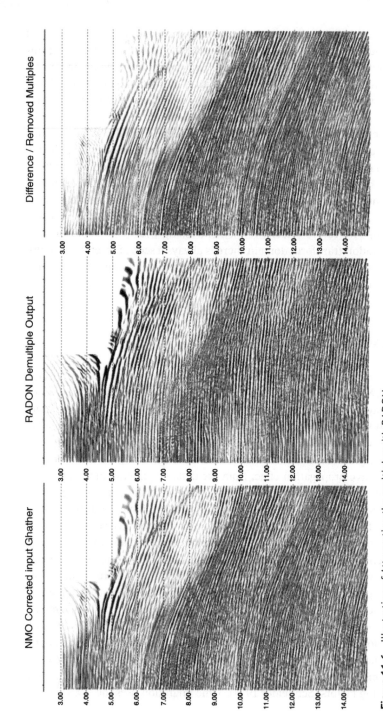

Figure 11.6 Illustration of Attenuating the multipleswith RADON.

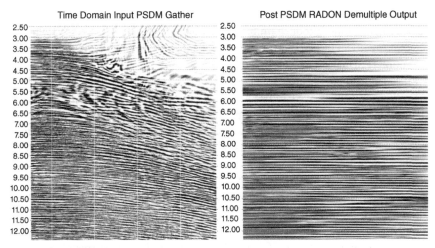

Figure 11.7 Illustration of the RADON demultiple technique after post migration to eliminate the effect of residual multiple reflections.

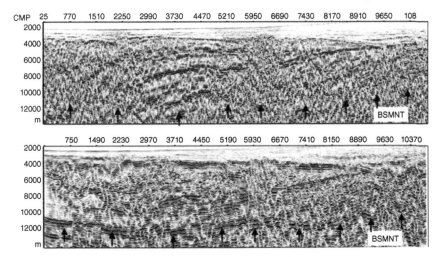

Figure 11.8 Representative seismic sections before (top) and after (bottom) multiple attenuation.

model as an input while depth migration uses an interval velocity, i.e. the actual subsurface velocity model. Using accurate velocity for pre-stack migration, we get a flat image gather. This allows us to use pre-stack migration as a velocity analysis tool to refine the input velocity in time or at the depth domain. The usual technique is to perturb the velocity field until a model obtained for which we get a flat pre-stack migrated common image point gathers. Unfortunately, an incorrect velocity model can also produce a good image with flat gathers (Stork 1992). For migration, if the velocity field lens's rays, cause multi-pathing, even a correct velocity field cannot produce uniformly flat gathers (Nolan and Symes 1996). To observe the effect of derived traveltime velocities, we have run Pre-Stack Depth migration (PSDM) along a seismic migration.

Effects of residual multiple reflection can still be seen in image gathers; to eliminate these multiples, we can use a narrow pass band for slowness in the radon Demultiple technique to get a multiple free image. As radon transform can only be performed in time domain, before performing radon Demultiple, the image gather was scaled into the time domain. The effect of post migration Demultiple can be seen in Figure 11.7.

Post stack data conditioning: After the post migration Demultiple in time domain data was stacked and some post stack conditionings such as FK filter, FX deconvolution, time varying filtering, and post stack mute are applied. Finally, the section is scaled to the depth domain to get the conditioned pre-stack depth migrated section (Figure 11.8) showing improved seismic sections to the PSDM in which we can easily interpret the boundaries such as a trap top and basement.

12

Future Scope

12.1 Introduction

Using active-source seismic tomography techniques, traveltime and full wave-form tomography, we aim to achieve different objectives contingent to the study area and to the chapters covered in both synthetic and field seismic data applications. We briefly summarize all topics for a generalized overview.

12.2 Limitations of the Study

Since we applied the full waveform tomography (FWT) in acoustic approximation, we excluded the deep reflections and converted phases. Although the acoustic approximation is one of the limitations of our FWT, it has an advantage on computational aspects. If we extend the study to elastic approximation without a matter of computational efficiency, then we have an opportunity to include all the phases similar to field seismic data so that we can get good control of the inversion algorithm and related results. Moreover, in our study, we didn't consider the anisotropic phenomena of the subsurface. This also one of the most important limitation of our study.

12.3 Future Research Scope

The velocity models can be used in migration to generate the best possible sections and may be converted to pore-pressure estimates using the rock physics. These studies are very useful in obtaining knowledge regarding faults and tectonics activities related to earthquakes. We did all of the work in acoustic approximation,

Active Seismic Tomography: Theory and Applications, First Edition. Kalachand Sain and Damodara Nara.
© 2023 John Wiley & Sons, Inc. Published 2023 by John Wiley & Sons, Inc.

which concentrates mostly P-wave information by excluding all other recorded information. Some of the specific things we need to understand during traveltime are full waveform tomography aspects mentioned in this list:

- Multi-parameter inversion is first and foremost the future work of the study by understanding the sensitivities of each parameter.
- Once we get through enough with simple acoustic approximations, we have the chance of extending this into elastic and anisotropic approximations.
- Over the advancement of the computational resources, 3D applications play a vital role in understanding the subsurface.
- We need to look in either designing acquisition geometries, which suits to full waveform tomography, or to advance the knowledge by which we can utilize the existing data sets efficiently to delineate the subsurface information.
- Although we studied FWT for the last two decades, still we are lagging in understanding problems such as cycle-skipping, low-frequency data recording, proper selection of damping parameter in complex frequency implementation, and proper selection the norms (l2, l1, logarithmic, Huber, correlation, etc.).

12.4   Concluding Remarks

Full waveform tomography is a state-of-the-art technique used for delineating the finer details in the sub-wavelength range and has very limited successful field seismic data applications. One success is that we have applied and presented the imaging results for better understanding of the sedimentary basins of the Indian sub-continent; however, we are limited to only acoustic approximations and only to limited data sets; this technique can be utilized for future data sets, and it can be extended to multi-parameter inversions. Further, these type of optimum sub-surface velocity models also useful in improving the seismic sections, which are able to develop from multi-channel seismic data by migration techniques.

References

Aki, K. and Lee, W.H.K. (1976). Determination of the three-dimensional velocity anomalies under a seismic array using first P arrival times from local earthquakes 1. A homogeneous initial model. *J. Geophysics Res.* 81: 4381–4399.

Aki, K. and Richards, P.G. (1980). *Quantitative Seismology - Theory and Methods*. San Francisco: W. H. Freeman and Company.

Alford, R.M., Kelly, K.R., and Boore, D.M. (1974). Accuracy of finite-difference modeling of the acoustic wave equation. *Geophysics* 39: 834–842.

Alkhalifah, T. (2015a). Scattering-angle based filtering of the waveform inversion gradients. *Geophysics J. Int.* 200: 363–373.

Alkhalifah, T. (2015b). Conditioning the full-waveform inversion gradient to welcome anisotropy. *Geophysics* 80: R111–R122.

Alterman, Z. and Karal, F.C. (1968). Propagation of elastic waves in layered media by finite-difference methods. *Bull. Seism. Soc. Am.* 58: 367–398.

Anstey, N.A. (1977). *Seismic Interpretation: The Physical Aspects*. Boston: IHRDC.

Bamberger, A., Chavent, G., Hemon, C., and Lailly, P. (1982). Inversion of normal incidence seismograms. *Geophysics* 47 (5): 757–770.

Barthelmes, F. and Köhler, W. (2016). International Centre for Global Earth Models (ICGEM). volume 90 of The Geodesists Handbook. 907–1205.

Ben-Hadj-Ali, H., Operto, S., and Virieux, J. (2008). Velocity model building by 3D frequency-domain, full-waveform inversion of wide-aperture seismic data. *Geophysics* 73 (5): VE101–VE117.

Berenger, J.P. (1994). A perfectly matched layer for absorption of electromagnetic waves. *J. Computers. Phys.* 114: 185–200.

Berkhout, A.J. and Verschuur, D.J. (1997). Estimation of multiple scattering by iterative inversion, Part1: theoretical consideration. *Geophysics* 62 (N5): 1596–1611.

Active Seismic Tomography: Theory and Applications, First Edition. Kalachand Sain and Damodara Nara.
© 2023 John Wiley & Sons, Inc. Published 2023 by John Wiley & Sons, Inc.

Biswas, S.K. and Singh, N.K. (1988). Western continental margin of India and hydrocarbon potential of deep-sea basins. Proceedings of the South East Asia Petroleum Exploration Society VIII: 100–113.

Bohlen, T. (1998). Interpretation of measured seismograms by means of visco-elastic finite difference modeling. Ph.D. thesis, Kiel University.

Bohlen, T. (2002). Parallel 3-D viscoelastic finite-difference seismic modeling. *Computers. and Geosciences.* 28 (8): 887–899.

Bois, P., La Porte, M., Lavergne, M., and Thomas, G. (1972). Well-to-well seismic measurements. *Geophysics* 37: 471–480.

Boore, D.M. (1970). Love waves in nonuniform waveguides: finite difference calculations. *J. Geophysics Res.* 1970: 1512–1527.

Brenders, A.J. and Pratt, R.G. (2007a). Full waveform tomography for lithospheric imaging: results from a blind test in a realistic crustal model. *Geophysics J. Int.* 168: 133–151.

Brenders, A.J. and Pratt, R.G. (2007b). Efficient waveform tomography for lithospheric imaging: implications for realistic 2D acquisition geometries and low frequency data. *Geophysics J. Int.* 168: 152–170.

Brossier, R., Operto, S., and Virieux, J. (2009). Seismic imaging of complex onshore structures by 2D elastic frequency-domain full-waveform inversion. *Geophysics* 74 (6): WCC105–WCC118.

Brossier, R. and Virieux, J. (2011). *Lecture Notes on Full Waveform Inversion. SEISCOPE Consortium.* Grenoble, France: Universite Joseph Fourier.

Carcione, J.M., Herman, G.C., and ten Kroode, A.P.E. (2002). Seismic modeling. *Geophysics* 67 (4): 1304–1325.

Cerjan, C., Kosloff, D., Kosloff, R., and Reshef, M. (1985). A nonreflecting boundary condition for discrete acoustic and elastic wave equations. *Geophysics* 50: 705–708.

Cerveny, V. (1987). Ray tracing algorithms in three-dimensional laterally varying layered structures. Seismic tomography: With applications in global seismology and exploration geophysics. D. Reidel and Dordrecht, D. Reidel and Dordrecht, 99–133.

Cerveny, V. (2001). *Seismic Ray Theory.* Cambridge: Cambridge University Press.

Chatterjee, S.M., Sar, D., Boruah, S., Nabakumar, K.H., and Sanjeev, S. (2006). Exploration leads from gravity and magnetic data in Kerala-Konkan basin, India. *SPG 6th Int. conf. & Exposition on petroleum Geophysics.* Kolkata. 66–71.

Clayton, R. and Engquist, B. (1977). Absorbing boundary conditions for acoustic and elastic wave equations. *Bull. Seismol. Soc. Am.* 67: 1529–1540.

Collino, F. and Tsogka, C. (2001). Application of the perfectly matched absorbing layer model to the linear elastodynamic problem in anisotropic heterogeneous media. *Geophysics* 66: 294–307.

Colombo, D. (2005). Benefits of wide-offset seismic for commercial exploration targets and implications for data analysis. *Lead. Edge.* 24: 352–363.

Damodara, N., Vijaya Rao, V., Kalachand, S., Prasad, A.S.S.S.R.S., and Murty, A.S.N. (2017). Basement configuration of the West Bengal sedimentary basin, India as revealed by seismic refraction tomography: its tectonic implications. *Geophysics J. Int.* 208: 1490–1507.

Dessa, J.X., Operto, S., Kodaira, S., Nakanishi, A., Pascal, G., Virieux, J., and Kaneda, Y. (2004). Multiscale seismic imaging of the eastern Nankai trough by full waveform inversion. *Geophysics Res. Lett.* 31: L18606.

Dobrin, M.B. and Savit, C.H. (1988). *Introduction to Geophysical Prospecting*, 4 Sub edition. Mcgraw-Hill College.

Eccles, J.D., White, R.S., and Christie, P.A.F. (2009). Identification and inversion of converted shear waves: case studies from the European North Atlantic continental margins. *Geophysics J. Int.* 179: 381–400.

Eccles, J.D., White, R.S., Robert, A.W., and Christie, P.A.F. (2007). Wide angle converted shear wave analysis of a north Atlantic volcanic rifted continental margin: constraint on sub-basalt lithology. *First Break* 25 (10): 63–70.

Firbas, P. (1981). Inversion of travel-time data for laterally heterogeneous velocity structure – linearization approach. *Geophys J. R. Astron. Soc.* 67: 189–198.

Gauthier, O., Virieux, J., and Tarantola, A. (1986). Two-dimensional non-linear inversion of seismic waveforms: numerical results. *Geophysics* 51: 1387–1403.

Geller, R.J. and Hara, T. (1993). Two efficient algorithms for iterative linearized inversion of seismic waveform data. *Geophysics J. Internat.* 115: 699–710.

Gombos, A.M., Jr., Powell, W.G., and Norton, I.O. (1995). The tectonic evolution of Western India and its impact on hydrocarbon occurrences: an overview. *Sediment. Geol.* 96: 119–129.

Gorszczyk, A., Operto, S., and Malinowski, M. (2017). Towards a robust workflow for deep crustal imaging by FWI of OBS data: the eastern Nankai trough revisited. *J. Geophysics Res.* 122 (6): 4601–4630.

Hammer, P.T.C., Dorman, L.M., Hildebrand, J.A., and Cornuelle, B.D. (1994). Jasper Seamount structure: seafloor seismic refraction tomography. *J. Geophysics Res.* 99: 6731–6752.

Hanssen, M. (2005). Processing techniques for wide-angle seismic data. PhD thesis, University of Cambridge, Wolfson College Cambridge.

Hardi, B.I. and Sanny, T.A. (2016). *Numerical Modeling: Seismic Wave Propagation in Elastic Media Using Finite-difference and Staggered-grid Scheme*. Indonesia: 41th HAGI Annual Convention and Exhibition.

Hobro, J.W.D., Minshull, T.A., Singh, S.C., and Chand, S. (2005). A three-dimensional seismic tomographic study of the gas hydrate stability zone, offshore Vancouver Island. *J. Geophysics Res.* 110: B09102. doi: 10.1029/2004JB003477.

Hobro, J.W.D., Singh, S.C., and Minshull, T.A. (2003). Three-dimensional tomographic inversion of combined reflection and refraction seismic traveltime data. *Geophysics J. Int.* 152: 79–93.

Hole, J.A. (1992). Nonlinear high-resolution three-dimensional seismic travel time tomography. *J. Geophys. Res.* 97: 6553–6562.

Hole, J.A. and Zelt, B.C. (1995). 3-D finite difference reflection traveltimes. *Geophysics J. Int.* 121: 427–434.

Huang, J. and Bellefleur, G. (2012). Joint transmission and reflection traveltime tomography using the fast sweeping method and the adjoint-state technique. *Geophysics J. Int.* 188 (2): 570–582. doi: https://doi.org/10.1111/j.1365-246X.2011.05273.x.

Hustedt, B., Operto, S., and Virieux, J. (2004). Mixed-grid and staggered-grid finite-difference methods for frequency-domain acoustic wave modeling. *Geophysics J. Int.* 157: 1269–1296.

Iturrarán-Viveros, U. and Sánchez-Sesma, F.J. (2011). Seismic wave propagation in real media: numerical modeling approaches. In *Encyclopedia of Solid Earth Geophysics. Encyclopedia of Earth Sciences Series* (ed. H.K. Gupta) Dordrecht: Springer. doi: https://doi.org/10.1007/978-90-481-8702-7_6.

Jones, G.D. (2003). Velocity images by stacking slowness-depth seismic wave fields. PhD thesis, University of Cambridge.

Kamei, R., Pratt, R.G., and Tsuji, T. (2013). On acoustic waveform tomography of wide-angle OBS data-strategies for pre-conditioning and inversion. *Geophysics J. Int.* 194: 1250–1280.

Kapoor, S., Vigh, D., Wiarda, E., and Alwon, S. (2013). Full waveform inversion around the world. In: *Conference Proceedings, 75th EAGE Conference & Exhibition incorporating SPE EUROPEC 2013*, June 2013, cp-348-00693. doi: 10.3997/2214-4609.20130827.

Kearey, P., Brooks, M., and Hill, I. (2002). *An Introduction to Geophysical Exploration*, 3e. Blackwell Science Ltd.

Kolb, P., Collino, F., and Lailly, P. (1986). Pre-stack inversion of a 1-D medium. *Proc. IEEE* 74: 498–508.

Korenaga, J., Holbrook, W.S., Kent, G.M., Kelemen, P.B., Detrick, R.S., Larsen, H.C., Hopper, J.R., and Dahl-Jensen, T. (2000). Crustal structure of the southeast Greenland margin from joint refraction and reflection seismic tomography. *J. Geophysics Res.* 105: 21,591–21,614.

Kurin, E., Kovozov, A., Muzichenko, E., and Rao, C.V. (2015). Multiple attenuation in deep-water settings complicated by basalt traps: a case study. *Society of Petroleum Geophysicists India, 11ᵗʰ Biennial international conference and exposition* Jaipur.

Kurzmann, A., Przebindowska, A., Kohn, D., and Bohlen, T. (2013). Acoustic full waveform tomography in the presence of attenuation: a sensitivity analysis. *Geophysics J. Int.* 195 (2): 985–1000.

Lailly, P. (1983). The seismic inverse problem as a sequence of before stack migrations. In: *Conference on Inverse Scattering: Theory and Application* (ed. J.B. Bednar, R. Redner, E. Robinson, and A. Weglein) Soc. Industr. Appl. Math.

Lavander, A., Zelt, C.A., and Symes, W.W. (2007). Active source studies of crust and lithosphere structure. volume 1 of Treatise on Geophysics, Seismology and structure of the Earth. 247–288.

Lelièvre, P.G., Farquharson, C.G., and Hurich, C.A. (2011). Inversion of first-arrival seismic traveltimes without rays, implemented on unstructured grids. *Geophysics J. Int.* 185 (2): 749–763. doi: https://doi.org/10.1111/j.1365-246X.2011.04964.x.

Lutter, W.J. and Nowack, R.L. (1990). Inversion for crustal structure using reflections from the PASSCAL Ouachita experiment. *J. Geophysics Res.* 95: 4633–4646.

Lutter, W.J., Nowack, R.L., and Braile, L.W. (1990). Seismic imaging of upper crustal structure using travel times from the PASSCAL Ouachita experiment. *J. Geophysics Res.* 95: 4621–4631.

Madariaga, R. (1976). Dynamics of an expanding circular fault. *Bull. Seism. Soc. Am.* 65: 163–182.

Malinowski, M. and Operto, S. (2008). Quantitative imaging of the Permo-Mesozoic complex and its basement by frequency domain waveform tomography of wide-aperture seismic data from the Polish Basin. *Geophysics Prospect.* 56: 805–825.

Marfurt, K.J. (1984). Accuracy of finite-difference and finite-element modeling of the scalar and elastic wave equations. *Geophysics* 49 (5): 533–549.

Maslen, G. (2013). *A Technical Blog about Seismic Data Processing*. GLOBE Claritas.

McCaughey, M. and Singh, S.C. (1997). Simultaneous velocity and interface tomography of normal-incidence and wide-aperture traveltime data. *Geophysics J. Int.* 131: 87–99.

Menke, W. (1989). Geophysical data analysis: discrete inverse theory. In *International Geophysics Series*, rev.ed, 45, San Diego, CA: Academic Press Inc.

Meyer, B., Saltus, R., and Chulliat, A. (2017). EMAG2: earth magnetic anomaly grid (2-arc-minute resolution) Version 3. National centers for environmental information, NOAA. Model. http://dx.doi.org/10.7289/V5H70CVX.

Moczo, P., Kristek, J., and Galis, M. (2014). *The Finite Difference Modelling of Earthquake Motions*. Cambridge: Cambridge University Press.

Moczo, P., Kristek, J., Galis, M., Pazak, P., and Balazovjech, M. (2007). The finite-difference and finite-element modeling of seismic wave propagation and earthquake motion. *Acta. Phys. Slovaca* 57 (2): 177–406.

Mora, P.R. (1987). Nonlinear two-dimensional elastic inversion of multi offset seismic data. *Geophysics* 52: 1211–1228.

Murty, A.S.N., Sain, K., Sridhar, V., Prasad, A.S.S.S.R.S., and Raju, S. (2016). Delineation of Trap and subtrappean Mesozoic sediments in Saurashtra peninsula. *Curr. Sci.* 110: 1844–1851.

Nara, D. and Sain, K. (2018). Workflow for an initial model of seismic tomography. 8th International Conference & Exhibition of EAGE Saint Petersburg, Russia.

Nolan, C.J. and Symes, W. (1996). Imaging and coherency in complex structures: 66th. *Ann. Mtg. Soc. Expl. Geophysics* Expanded Abstracts: 359–362. https://doi.org/10.1190/1.1826642.

Nolet, G. (1987). Seismic wave propagation and seismic tomography. In: *Seismic Tomography. Seismology and Exploration Geophysics*, vol. 5 (ed. G. Nolet). Dordrecht: Springer. doi: 10.1007/978-94-009-3899-1_1.

Norton, I.O. and Sclater., J.G. (1979). A model for the evolution of indian ocean and breakup of Gondwanaland. *J. Geophysics Res.* 84 (B12): 6803–6830.

Nowack, R. and Braile, L. (1993). Refraction and wide-angle reflection tomography: theory and results. Seismic tomography: Theory and practice, 733–765. London: Chapman and Hall publishers.

Operto, S., Ravaut, C., Improta, L., Virieux, J., Herrero, A., and Dell'Aversana, P. (2004). Quantitative imaging of complex structures from dense wide-aperture seismic data by multiscale traveltime and waveform inversions: a case study. *Geophysics Prospect.* 52: 625–651.

Operto, S., Virieux, J., Dessa, J.X., and Pascal, G. (2006). Crustal seismic imaging from multifold ocean bottom seismometer data by frequency-domain full-waveform tomography: application to the eastern-Nankai trough. *J. Geophysics Res.* 111: B09306.

Prasad, A.S.S.S.R.S., Sain, K., and Sen, M.K. (2013). Imaging sub basalt Mesozoics along JakhauMandvi and Mandvi-Mundra profiles in Kutch sedimentary basin from seismic and gravity modelling. *Geohorizons* 18 (2): 51–56.

Pratt, R.G. (1990a). Frequency-domain elastic wave modeling by finite differences: a tool for crosshole seismic imaging (short note). *Geophysics* 55: 626–632.

Pratt, R.G. (1990b). Inverse theory applied to multi-source cross-hole tomography II: elastic wave-equation method. *Geophysics Prosp.* 38: 311–330.

Pratt, R.G. (1999). Seismic waveform inversion in the frequency domain, Part 1: theory and verification in physical scale model. *Geophysics* 64: 888–901.

Pratt, R.G. and Goulty, N.R. (1991). Combining wave-equation imaging with traveltime tomography to form high-resolution images from crosshole data. *Geophysics* 56: 208–224.

Pratt, R.G., Shin, C., and Hicks, G.J. (1998). Gauss-Newton and full Newton methods in frequency-space seismic waveform inversion. *Geophysics J. Int.* 133: 341–362.

Pratt, R.G. and Shipp, R.M. (1999). Seismic waveform inversion in the frequency domain. Part 2. Fault delineation in sediments using crosshole data. *Geophysics* 64: 902–914.

Pratt, R.G. and Worthington, M.H. (1988). The application of diffraction tomography to cross-hole seismic data. *Geophysics* 53: 1284–1294.

Pratt, R.G. and Worthington, M.H. (1990). Inverse theory applied to multi-source cross-hole tomography I: acoustic wave-equation method. *Geophysics Prosp.* 38: 287–310.

Qin, F., Luo, Y., Olsen, K.B., Cai, W., and Schuster, G.T. (1992). Finite-difference solution of the eikonal equation along expanding wavefronts. *Geophysics* 57 (3): 478–487.

Ravaut, C., Operto, S., Improta, L., Virieux, J., Herrero, A., and Dell'Aversana, P. (2004). Multiscale imaging of complex structures from multifold wide-aperture seismic data by frequency-domain full-waveform tomography: application to a thrust belt. *Geophysics J. Int.* 159: 1032–1056.

Rawlinson, N. and Sambridge, M. (2003). Seismic Traveltime Tomography of the Crust and Lithosphere. *Advances in Geophysics* 46: 81–198.

Robein, E. (2003). Velocities, Time-imaging, and Depth-imaging in Reflection Seismics: Principles and Methods. *Eur. Assn. Geosciences. Eng. Publications.*

Sain, K., Bruguier, N., Murthy, A.S.N., and Reddy, P.R. (2000). Shallow velocity structure along the Hirapur-Mandla profile in central India, using travel time inversion of wide-angle seismic data, and its tectonic implications. *Geophysics J. Int.* 142: 505–515.

Sain, K. and Kaila, K.L. (1996). Ambiguity in the solution of the velocity inversion problem and a solution by joint inversion of seismic refraction and wide-angle reflection times. *Geophysics Jour. Int.* 124: 215–227.

Sain, K., Nara, D., Chandrasekhar, N., Pandurangi, S., and Katiyar, G. (2017). Imaging finer details of subsurface from seismic full waveform tomography and its application to field seismic data in Kerala - Konkan offshore. In 12th Biennial International conference and exposition, Jaipur, Rajasthan, India.

Sain, K., Reddy, P.R., and Behera, L. (2002b). Imaging of low-velocity Gondwana sediments in the Mahanadi delta of India using travel time inversion of first arrival seismic data. *J. App. Geophysics* 49: 163–171.3.

Sain, K., Zelt, C.A., and Reddy, P.R. (2002a). Imaging subvolcanic Mesozoics using travel time inversion of wide-angle seismic data in the Saurashtra peninsula of India. *Geophysics J. Int.* 150: 820–826.

Scales, J.A., Dochety, P., and Gerstenkorn, A. (1990). Regularization of nonlinear inverse problems: imaging the near surface weathering layer. *Inverse Problems* 6: 115–131.

Schneider, W.A. (February 1978). Integral formulation for migration in two and three dimensions. *Geophysics* 43: 49–76.

Sen, M.K. and Stoffa, P.L. (2013). *Global Optimization Methods in Geophysical Inversion*, 2e. UK: Cambridge University Press.

Sheriff, R.E. (2002). *Encyclopaedic Dictionary of Exploration Geophysics*, 4e. Society of Exploration Geophysicists.

Sheriff, R.E. and Geldart, L.P. (1995). *Exploration Seismology*, 2e Cambridge University Press.

Shin, C. and Cha, Y.H. (2008). Waveform inversion in the Laplace domain. *Geophysics J. Int.* 173: 922–931.

Shin, C.S. (1988). Non-linear elastic wave inversion by blocky parameterization. Ph.D. thesis, Univ. of Tulsa.

Sibuet, J.-C., Klingelhoefer, F., Huang, Y.-P., Yeh, Y.-C., Rangin, C., Lee, C.-S., and Hsu, S.-K. (2016). Thinned continental crust intruded by volcanics beneath the northern Bay of Bengal. *Mar. Petro. Geol.* 77: 471–486.

Singh, N.K. and Lal, N.K. (1993). Geology and petroleum prospects of Konkan – kerala basin. Proc. 2nd Seminar on Petroleum Basins of India, 2. 461–469. KDMIPE and ONGC.

Sirgue, L. and Pratt, R.G. (2004). Efficient waveform inversion and imaging: a strategy for selecting temporal frequencies. *Geophysics* 69: 231–248.

Song, Z.M., Williamson, P.R., and Pratt, R.G. (1995). Frequency-domain acoustic-wave modeling and inversion of crosshole data II-Inversion method synthetic experiments and real-data results. *Geophysics* 60: 796–809.

Sourbier, F., Operto, S., Virieux, J., Amestoy, P., and L'Excellent, J. (2009). FWT2D: a massively parallel program for frequency-domain full-waveform tomography of wide-aperture seismic data—Part 2 Numerical examples and scalability analysis. *Computers. and Geosciences.* 35: 496–514.

Spence, G.D., Clowes, R.M., and Ellis, R.M. (1985). Seismic structure across the active subduction zone of western Canada. *J. Geophysics Res.* 90: 6754–6772.

Sridhar, A.R., Prasad, A.S.S.S.R.S., Satyavani, N., and Sain, K. (2009). Sub-Trappean Mesozoic sediments in the Narmada basin based on traveltime and amplitude modeling - a revisit to old seismic data. *Curr. Sci.* 97: 1462–1466.

Stork, C. (1992). Reflection tomography in the post migrated domain. *Geophysics* 57: 680–692.

Strang, G. and Fix, G.J. (1988). *An Analysis of the Finite Element Method*. Cambridge: Wellesley Cambridge Press.

Taillandier, C., Noble, M., Chauris, H., and Calandra, H. (2009). First-arrival traveltime tomography based on the adjoint-state method. GEOPHYSICS 74: WCB1-WCB10.

Tarantola, A. (1984). Inversion of seismic reflection data in the acoustic approximation. *Geophysics* 49: 1259–1266.

Thurber, C.H. (1983). Earthquake locations and three-dimensional crustal structure in the Coyote Lake area, central California. *J. Geophysics Res.* 88: 8226–8236.

Toomey, D.R., Solomon, S.C., and Purdy, G.M. (1994). Tomographic imaging of the shallow crustal structure of the East Pacific Rise at 930'N. *J. Geophysics Res.* 99: 24,135–24,157.

Van Avendonk, H.J.A., Harding, A.J., Orcutt, J.A., and McClain, J.S. (1998). A two-dimensional tomographic study of the Clipperton transform, fault. *J. Geophysics Res.* 103: 17,885–17,899.

Vidale, J.E. (1988). Finite difference calculation of traveltimes. *Bull. Seism. Soc. Am.* 78: 2062–2076.

Vidale, J.E. (1990). Finite difference calculation of traveltimes in three dimensions. *Geophysics* 55: 521–526.

Vigh, D. and Starr, E.W. (2008). 3D prestack plane-wave, full-waveform inversion. *Geophysics* 73 (5): VE135–VE144.

Vijaya Rao, V., Damodara, N., Sain, K., Sen, M.K., Murty, A.S.N., and Sarkar, D. (2015). Upper crust of the Archean Dharwar Craton in southern India using seismic refraction tomography and its geotectonic implications. *Geophysics J. Int.* 200: 652–663.

Virieux, J. and Madariaga, R. (1982). Dynamic faulting studied by a finite difference method. *Bull. Seism. Soc. Am.* 72: 345–369.

Virieux, J. and Operto, S. (2009). An overview of full waveform inversion in exploration geophysics. *Geophysics* 74: WCC127–WCC152.

Waheed, U. and Alkhalifah, T. (2015). Fast sweeping methods for factored TTI eikonal equation. 77th Annual International Conference and Exhibition, EAGE, Extended abstracts. doi: 10.3997/14*4609.201412998.

Waheed, U., Alkhalifah, T., and Wang, H. (2015). Efficient traveltime solutions of the acoustic TI eikonal equation. *J. Computers. Phys.* 282: 62–76.

Waheed, U., Flagg, G., and Yarman, C.E. (2016). First-arrival traveltime tomography for anisotropic media using the adjoint-state method. *GEOPHYSICS* 81: R147–R155.

Waheed, U., Yarman, C.E., and Flagg, G. (2014). An iterative fast sweeping based eikonal solver for tilted orthorhomic media: 84th annual international meeting, SEG, Expanded abstracts 480–485.

Wang, Y. and Pratt, R.G. (1997). Sensitivities of seismic traveltimes and amplitudes in reflection tomography. *Geophysics J. Int.* 131: 618–642.

Wang, Y. and Rao, Y. (2009). Reflection seismic waveform tomography. *J. Geophysics Res.* 114: B03304.

Wessel, P., Smith, W.H.F., Scharroo, R., Luis, J. and Wobbe, F. (2013). Generic Mapping Tools: Improved Version Released, Eos Trans. AGU, 94 (45): 409.

White, D.J. (1989). Two-dimensional seismic refraction tomography. *Geophysics J.* 97: 223–245.

Wiggins, J.W. (1988). Attenuation of complex water bottom multiples by wave equation based prediction and subtraction. *Geophysics* 53: 1527–1539.

Williamson, P.R. (1990). Tomographic inversion in reflection seismology. *Geophysics J. Int.* 100: 255–274.

Worthington, M.H. (1984). An introduction to seismic tomography. *First Break* 2 (11): 20–26.

Wu, R.S. and Toksöz, M.N. (1987). Diffraction tomography and multisource holography applied to seismic imaging. *Geophysics* 52: 11–25.

Yilmaz, Ö. (2001). *Seismic Data Analysis*, Investigations in Geophysics No. 10, vol. 1 Society of Exploration Geophysicists.

Zelt, C.A. (1998a). *FAST - Program Package for First Arrival Seismic Tomography.* Rice University.

Zelt, C.A. (1998b). Lateral velocity resolution from three-dimensional seismic refraction data. *Geophysics J. Int.* 135: 1101–1112.

Zelt, C.A. (1999). Modeling strategies and model assessment for wide-angle seismic traveltime data. *Geophysics J. Int.* 139: 183–204.

Zelt, C.A. and Barton, P.J. (1998). Three-dimensional seismic refraction tomography: a comparison of two methods applied to data from the Faeroe Basin. *J. Geophysics Res.* 103: 7187–7210.

Zelt, C.A., Sain, K., Naumenko, J.V., and Sawyer, D.S. (2003). Assessment of crustal velocity models using seismic refraction and reflection tomography. *Geophysics J. Int.* 153: 609–626.

Zelt, C.A. and Smith, R.B. (1992). Seismic traveltime inversion for 2-D crustal velocity structure. *Geophysics J. Int.* 108 (1): 16–34.

Zelt, C.A. and White, D.J. (1995). Crustal structure and tectonics of the southeastern Canadian Cordillera. *J. Geophysics Res.* 100: 24 255–24 273.

Zienkiewicz, O.C. and Taylor, R.L. (1989). *The Finite Element Method*, 4e, 1. New York: McGraw-Hill.

Zienkiewicz, O.C., Taylor, R.L., and Zhu, J.Z. (2013). *The Finite Element Method: Its Basis and Fundamentals*, 7e, 1. Cambridge: Elsevier.

Ziolkowski, A., Hanssen, P., Gatliff, R., Jakubowicz, H., Dobson, A., Hampson, G., Li, X.-Y., and Liu, E. (2003). Use of low frequencies for sub-basalt imaging. *Geophysics Prosp.* 51: 169–182.

Index

Active Seismic Tomography: Theory and Applications, First Edition. Kalachand Sain and Damodara Nara.
© 2023 John Wiley & Sons, Inc. Published 2023 by John Wiley & Sons, Inc.

Printed and bound by CPI Group (UK) Ltd, Croydon, CR0 4YY

16/04/2025

14658350-0001